STEM

STUDENT RESEARCH
HANDBOOK

STEM

STUDENT RESEARCH HANDBOOK

Darci J. Harland

National Science Teachers Association

Arlington, Virginia

Claire Reinburg, Director
Jennifer Horak, Managing Editor
Andrew Cooke, Senior Editor
Judy Cusick, Senior Editor
Wendy Rubin, Associate Editor
Amy America, Book Acquisitions
 Coordinator

ART AND DESIGN
Will Thomas Jr., Director
Joseph Butera, Senior Graphic Designer,
 Cover and Interior Design
Mary McCubbins, Group Icon Illustrator

PRINTING AND PRODUCTION
Catherine Lorrain, Director

NATIONAL SCIENCE TEACHERS ASSOCIATION
Francis Q. Eberle, PhD, Executive Director
David Beacom, Publisher
1840 Wilson Blvd., Arlington, VA 22201
www.nsta.org/store
For customer service inquiries, please call 800-277-5300.

NSTA is committed to publishing material that promotes the best in inquiry-based science education. However, conditions of actual use may vary, and the safety procedures and practices described in this book are intended to serve only as a guide. Additional precautionary measures may be required. NSTA and the authors do not warrant or represent that the procedures and practices in this book meet any safety code or standard of federal, state, or local regulations. NSTA and the authors disclaim any liability for personal injury or damage to property arising out of or relating to the use of this book, including any of the recommendations, instructions, or materials contained therein.

Library of Congress Cataloging-in-Publication Data
Harland, Darci J., 1972-
 STEM student research handbook / by Darci J. Harland.
  p. cm.
 Includes bibliographical references and index.
 ISBN 978-1-936137-24-4 (alk. paper)
 1. Research--Study and teaching (Secondary) I. Title.
 Q180.A1H37 2011
 507.1'2--dc23
                    2011023245

eISBN 978-1-936137-41-1

CONTENTS

CHAPTER 1
Beginning a STEM Research Project

CHAPTER 2
Research Design

CONTENTS

CHAPTER 3
Background Research and Note Taking

CHAPTER 4
Writing Hypotheses

CONTENTS

CHAPTER 5
Proposal Writing

CHAPTER 6
Organizing a Laboratory Notebook

CHAPTER 7
Descriptive Statistics

CONTENTS

CHAPTER 8
Graphical Representations

CHAPTER 9
Inferential Statistics and Data Interpretation

CHAPTER 10
Documentation and Research Paper Setup

CONTENTS

CHAPTER 11
Writing the STEM Research Paper

CHAPTER 12
Presenting the STEM Research Project

DEDICATION

This book is dedicated to all my former students. I am indebted to each of you. Your willingness to work hard, with minimal grumbling, has been an inspiration. My goal was never to turn each of you into STEM researchers but rather to help you gain a clear understanding of how scientific knowledge is acquired. I thank you for putting up with my inadequacies and my pat sayings and for trusting me when I told you the hard work would pay off.

PREFACE

Our nation's success depends on strengthening America's role as the world's engine of discovery and innovation. ... CEOs...understand that their company's future depends on their ability to harness the creativity and dynamism and insight of a new generation. And that leadership tomorrow depends on how we educate our students today—especially in science, technology, engineering and math.

—President Barack Obama, September 16, 2010 (Sabochik 2010)

The importance of improving science, technology, engineering, and mathematics (STEM) education has become a popular topic in recent years. It is clear, however, that the STEM education that today's high school students receive rarely mirrors what individuals in STEM careers actually do. Students are focused more on memorization than on identifying problems and finding ways to solve those problems. *STEM Student Research Handbook* engages students with the same inquiry skills used by STEM professionals. The handbook supports students as they practice skills of designing and conducting experiments and analyzing and presenting their findings.

I believe that the primary reason STEM educators do not include student-directed research as part of their curriculum is that they themselves have limited experience in this area. My goal in writing this book was to provide a practical resource that teachers and students can use to become actively engaged with topics that interest them as they are guided through the stages of a long-term research project. I hope that this handbook bridges a gap between STEM professionals and classroom teachers by providing a resource that will help students experience learning in the way scientists do, by doing research.

My experience is similar to that of many other teachers who at one point decide to implement a research component in their classrooms. Having designed and completed only one science experiment on my own as an undergraduate and only educational research as a graduate student, I was uncertain about my ability to lead high school students in performing research. When I received the first set of student research papers, it was obvious that I had to make changes in how I supported students through the research process. Over the years, as I identified resources, modified deadlines, and developed activities that helped students to implement the scientific process for themselves, I saw student research papers improve greatly. This handbook is a compilation of years of work, lessons learned from mistakes, and the good advice of other teachers and STEM professionals.

The handbook addresses the two major aspects of conducting research: planning and conducting experiments and then analyzing and communicating results through writing. First, the handbook provides a structure for STEM teachers to use as they work through the stages of the research process with their students. There is enough detail here so that even teachers who have never designed an experiment on their own can feel comfortable guiding students through theirs. Second, large segments of the handbook address the writing (aka language arts) issues involved with research. I believe that STEM teachers who are not specially trained in writing will find the writing instructions here to be invaluable. As a science teacher who has also studied and taught English, I have been able to help STEM students see the importance of communicating their results through effective writing. Whole chapters in this book cover note-taking techniques, proper documentation of research papers, and presentation preparation. Sadly, in-depth writing advice is commonly missing from other books teaching students how to conduct research.

The opportunity to design and conduct an experiment and then present the findings at local symposia has changed the lives of many of my high school students. They learned to take the initiative for their own learning, and many chose to enter STEM-related fields when they went to college. As I got better at leading my students through STEM research projects that they were really interested in, they started winning awards at the competitions, bringing home top prizes. In 2004, I was awarded the Sigma Xi (Illinois State University/ Illinois Wesleyan University chapter) Outstanding Science Teacher Award for my students' contributions to these competitions and for encouraging other local teachers to implement long-term research projects. Most rewarding to me, however, has been that my students actually performed the exciting research process from beginning to end and acquired better understandings of how STEM knowledge is advanced. That experience is completely different from listening to a lecture, observing a STEM professional on the job, or even performing hands-on labs that someone else designed.

Reference

Sabochik, K. 2010. Changing the equation in STEM education. The Whitehouse Blog: *www.whitehouse.gov/blog/2010/09/16/changing-equation-stem-education.*

ABOUT THE AUTHOR

Darci J. Harland, PhD, a former biology and English teacher, is the assistant director of research at the Center for Mathematics, Science, and Technology (CeMaST) at Illinois State University. Her educational experiences range from undergraduate and graduate education and biology courses, to high school and middle school science and English courses. Her research interests include the long-term impact of scientific research performed by high school and undergraduate students, the influence of personality on online and face-to-face classroom participation, the use of digital media as a tool for reflection, and the level of inquiry in one-to-one laptop classrooms.

ACKNOWLEDGMENTS

I would like to thank a number of individuals for the support and assistance they provided through the evolution of my life as well as the evolution of the *STEM Student Research Handbook*.

First of all, I am grateful to my parents, Craighton and Linda Hippenhammer. I am extremely appreciative for having been raised in a home that cultivated a wonder for the amazing world we live in and encouraged me to learn as much about it as possible. I am thankful for my husband, Craig, and my two boys who supported me during the long hours of writing.

A special thanks to the staff at the Center for Mathematics, Science, and Technology (CeMaST) at Illinois State University: Dr. William Hunter, whose encouraging words kept me writing when I wanted to quit; Ydalisse Pérez, Nicole Enzinger, and Dr. Jeff Helms, who contributed writing and conceptual support in areas where I am weak; Mary McCubbins for her expertise in graphic design; and Sara McCubbins and Amanda Fain for their editing.

Thanks also to Dr. Randall Johnson at Olivet Nazarene University, who modeled natural inquisitiveness whenever he asked, "How does it know?" I thank him for providing my first opportunity to conduct STEM research as an undergraduate in his ecology course. I now have an appreciation for the restraint he displayed when he answered my questions with more questions. He is an exemplary teacher.

Last, but not least, I am grateful to Suzanne McGroarty, the veteran teacher whose shoes I attempted to fill after her retirement. She inspired this book! I so appreciate "the McGroarty legacy" and the opportunities I had to improve myself as a teacher because of it. Sue's love of science, genuine concern for her students, and positive attitude as a career-long teacher truly have made her an inspiration and model to STEM teachers everywhere.

INTRODUCTION

To the Teacher

If you are a high school STEM or other kind of science teacher, you most likely already understand the value of having students *do science*. However, even with our best efforts to include inquiry in our teaching, the logistics of organizing learning experiences that encourage students to ask questions that they themselves answer is overwhelming. If you have considered, or are considering, implementing student-centered, long-term research projects, my guess is you have a few questions, such as, How do I monitor students doing various projects, at various stages? What if students want to design an experiment on a topic I know nothing about? What if the students get in over their heads? Can I include long-term research projects as part of my curriculum and still cover the required content? How can I ask my students to design an experiment if I have never done one? Questions like these are enough to send any teacher into a tailspin and to drop the idea forever. However, I am here to tell you, it can be done, and this handbook will show you how.

Incorporating Research Into the High School Science Curriculum

There are several ways that teachers, departments, and schools incorporate long-term research projects into the science curriculum. Sometimes it starts with a single teacher helping a few ambitious students who perform their own research before and after school. Other teachers add a long-term research project to an existing course and then work diligently to balance the content they need to address with the support students need to complete their projects. Some science departments systematically include research components throughout the curriculum, so that upon graduating, students have conducted multiple research projects at varying levels of difficulty. Sometimes a school is able to dedicate an entire course to student-centered research or provide a similar experience in summer enrichment opportunities. This handbook was written for any of the above scenarios.

Whatever your situation, I advise you to first decide how much time you want to dedicate to a research project and then decide on deadlines. As with any other unit plan, start with the end in mind. When do you want to have the research project (either a paper or poster presentation) in your

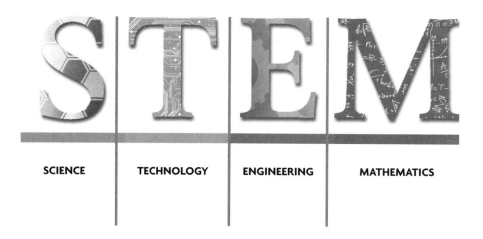

hands to grade? Once you have determined the final deadline, set deadlines to assess your students along the way. This handbook provides support in the construction of deadlines on several fronts. Appendix A is a sample checklist for developing deadlines. More indirectly, there are cues for you throughout the handbook. Although I wrote the book in language directed to your high school students, you will find references to "your teacher" throughout. I intended these as cues for you. I use phrases such as, "Your teacher may ask to see…" or "Your teacher will prefer either ___ or ___" as a prompt for you to discuss your requirements with your students. These cues suggest options of what sort of assessments to make along the way. I suggest you give grades throughout the length of the project, both formally and informally, to foster the concept of a journey of research, rather than giving a single grade to a final product.

Next you will need to decide how you want students to work—individually, in pairs, or in groups. On one hand, having students work individually simplifies the research process because each student performs each stage of the process individually and can choose a topic that is personally meaningful. For you, of course, individual projects will increase the total number of projects you must monitor and assess. On the other hand, allowing students to work in groups reduces your grading, but it does introduce other challenges—for example, students will often need guidance from you on how to divide the workload. (Throughout the handbook, a group icon—see p. xxvi—will signal tips for how groups might work together to accomplish the task at hand.) If you do choose the group route, I suggest that throughout the process you provide time for group members to frankly discuss their strengths and weaknesses, evaluate themselves and each other, and assign specific tasks to each individual in a signed contract that you also sign. Each group contract should cover the background research (Chapter 3), proposal (Chapter 5), data collection (Chapter 6), paper writing (Chapter 11), and presentation (Chapter 12) stages of the research project.

Using Outside Mentors

You may want to consider encouraging or even requiring your students to find a STEM mentor within the research field that interests them. Even if you have no local university or STEM industry companies in the vicinity, students can search online for possible mentors and be mentored at a distance. Your role when students work with a mentor is that of coach: You ensure that students meet deadlines, conduct the scientific process themselves (as much as possible), and communicate with their mentors. Support from mentors in the field frees you from having to be a content expert on each of the student projects. However, you are also releasing control of the level at which students experience the scientific process, particularly if a student is physically working in their mentor's lab. It is possible that the mentor's research interests will truncate your student's interest in the topic or field of study. Students may not have the opportunity to develop their own hypotheses and research designs but instead will participate in research currently being completed. The experience would still be a rich one, and beneficial to students, but it will differ from the experiences of students who do not have mentors.

Students who find online mentors are more likely to develop their own experimental design, using their mentor as a content expert and someone who can help them determine an appropriate research design for what they want to study. These mentors—if they become invested and prove to be reliable—are invaluable and provide students with an understanding of the research process that they may not otherwise receive. In this handbook, I refer students to "your teacher," but in the section of this introduction called "To the High School Student" (which begins on p. xxv), I let them know that if they are working with mentors, some of the references to *teacher* may actually refer to their *mentors*. You will need to communicate clearly with students regarding the differing roles between you and their mentor.

Do not underestimate your ability to coach students through research projects without mentors. Students can have successful research experiences with you as their primary resource. Even if you have never conducted a research project from the planning to the presentation phase, the *STEM Student Research Handbook* contains the details to guide you and your students comfortably through the process.

STEM Writing

After your students complete their research project, most likely you will ask them to either design a poster displaying their research results or to write a paper. In either case, students will be writing to communicate their results. It is important that you do not give short shrift to the writing part of their projects. Although it is natural that teachers trained in STEM subjects would be more interested in the experimental techniques and experiences than in the final paper and the writing steps that lead up to it, *it is crucial that you allow the necessary time for students to take notes, write up their designs and results, and write the final papers in preparation for presenting their experiments to an audience.* This handbook gives you great support in these areas (see especially Chapters 3, 10, and 11).

When I was in high school, I learned how and where to find resources and use them to support my thoughts or ideas. This really helped me when I was in college and writing research papers.

—Student Researcher

It would also be a good idea to talk to members of your English department about the writing aspect of your research project. Ask them at what level they require students to write their first large report and how they teach the report-writing process. This information will help you determine how much help your students will need. You are most likely to receive support from your English teachers if you use the note-taking strategies and documentation style that they teach. Most high school students will have written reports prior to your class, but it is possible that the idea of keeping detailed, organized notes to be used in a paper with reference citations is a new concept to them. Perhaps they have not, up until now, had to use parenthetical citations within a paper and have simply listed references haphazardly at the end of their papers. You will do your students a huge service by helping them understand the importance of competent documentation.

The most important lesson I learned in completing a research project was the basic research skills such as using databases, taking notes, and using citations properly.

—Student Researcher

I chose the MLA (Modern Language Association) documentation style as the one for students to use. MLA style is what most high school teachers use with their students, and it is the first documentation style they are likely to encounter in college. I completely understand a STEM teacher's resistance to MLA style since scientific papers are never written in this style. However, it is more important for students at this stage to understand the principles behind documentation. If you plan on having your students present their research at local, state, or national fairs or symposium contests, be sure to refer to their guidelines regarding documentation style. Most competitions do not require a specific style, only that it is applied consistently and correctly.

Another significant skill students should learn while doing this research project is how to do quality background research both online and at the library. I suggest that you contact your school and local librarians for help in organizing resources for your students. Although your school and local libraries may not have current paper resources on particular STEM topics, librarians can offer a session with students to help orient them to databases the school has access to along with any interlibrary-loan agreements that are available to students. (The technology icon—see p. xxvi—is used throughout the book to highlight tips for using technology during the research process.) I also highly recommend that, if possible, you organize a full-day field trip to a large university library, particularly if students are not introduced to a university library as part of their English courses. The greater number of STEM scholarly resources available through academic libraries will be worth the effort.

The Proposal Process

The first year I implemented research projects it became clear just how little students knew about applying the scientific method to their own experiments. My students could define the various aspects of an experiment, such as control, experimental groups, extraneous variables, and constants, but when it came to applying these concepts to their own research, they really struggled. I discovered that students need a lot of support in developing their research design. That's when I developed a proposal approval process (see Chapter 5). The proposal itself is quite an accomplishment for students, and the *STEM Student Research Handbook* supports them as they take small steps to reach a point where they can write a full research proposal. These steps include identifying questions they have about a topic, identifying possible independent and dependent variables, researching ways in which connections have already been made between them, and then writing a hypothesis to test their idea. The research design table in Chapter 2 will help students hone their ideas further. I encourage you to spend significant time on writing hypotheses and to preapprove them before students begin writing their proposals. Students' first significant grade should be the hypothesis.

I don't know if I ever really had an accurate understanding of the scientific research process before I did my own research project. I probably could've listed the steps for you, but until I actually did it myself I never really understood what it means in real life.

—Student Researcher

Completing the research project was the first time I ever really learned on my own. The teacher wasn't putting the information in front of me to memorize, rather I had to do my own research and how much I learned was directly related to how much effort I put into the research.

—Student Researcher

I also strongly suggest that you have students revise their proposals until you are confident that the proposals show that students have thoroughly researched the topic, that they have accounted for extraneous variables, and that their research designs are detailed enough to convince you that the students have a good chance of being successful. I call this type of assessment "Do Until Accepted" (DUA) (see Appendix A). To make this work, I have two due dates. The first is when students are required to turn in their first draft and the second is a week or two later. On each of these drafts, I write comments to help students improve their research designs. I don't give students an actual grade (in my grade book) until I give my approval for them to begin their research. If they want to receive an A on this assignment, they have to meet both deadlines. Within this time span, they can rewrite the proposal as many times as it takes to get it accepted. Some students will rewrite it three times and others nine times. Students who missed one of these deadlines cannot receive higher than a B on the proposal.

Research Symposia and Science Fairs

I encourage you to seek out an opportunity for your students to share their research at a research symposia or science fair. Knowing that individuals other than their teacher will be viewing and assessing their work is a strong motivator for students. You can locate competitions easily by searching online. Even if you decide not to attend someone else's event, I highly suggest you have an open house one evening where students showcase their research to parents, administrators, and community members. You could choose to have judges or just allow individuals to visit and talk with your student researchers about their projects.

I was never one to put more into school work than what needed to be, but because I was doing the research project on my own, and we were taking it to the symposium, I was more interested in the work that I was doing. I cared more about what the outcome would be instead of looking at how much work was put into it.

—Student Researcher

My last piece of advice is to pay attention to the balance between how much control you have over student projects and how much choice you give to students. Although students need structure, feedback, and support, it is also important that they have ownership of their projects. This may mean that students will choose a topic with which you are not familiar. I encourage you to allow students to include integrated STEM projects, even if their choices make you uncomfortable. For example, a student may have learned how to use a specific piece of equipment in another STEM course and wants to use it as part of this project. Admit your vulnerability, and agree to learn along with your student. Guiding students through the research process can be the most rewarding aspect of teaching.

All too often in college and high school, students just regurgitate the knowledge of others over and over in papers and projects. But this, an actual research project, forced me to come up with my own ideas for an experiment and formulate my own educated conclusions with the support of other research.

—Student Researcher

Importance of Student-Centered STEM Research

The scientific method is a common introductory topic within all science curricula (Bereiter and Scardamalia 2009). However, it is well documented that just because students can describe the scientific process doesn't mean that they are able to perform scientific thinking or show productive inquiry skills (Ayers and Ayers 2007; Leonard and Chandler 2003; Tang et al. 2010). Therefore, without having gone through the scientific process themselves from beginning to end, students are unlikely to truly understand the nature of science, especially that the process of scientific inquiry is often nonlinear.

Authentic research experiences have the potential to provide high school students with the scientific reasoning skills desired by both high school and university instructors. Although some STEM classrooms use labs with procedures where students simply record the results, others use inquiry and problem-based learning (PBL). Research done in K–12 classroom shows that when teachers implement problem-based projects and inquiry-based labs, students not only learn the same content as in lecture-based units but also gain critical thinking and problem-solving skills (Drake and Long 2009; Tarhan et al. 2008; Wong and Day 2009).

PBL and inquiry should have an important role in STEM courses. Unfortunately, the common model is still teacher-centered (Taraban et al. 2006)—the teacher decides the topic that students will study and the teacher sets up the problem or question that students will answer. Although students may be engaged and learning about problem solving, they are not designing their own experiments to address problems that they themselves have identified. When students are in control of their own research, it increases motivation and creates a strong sense of ownership (Marcus et al. 2010).

Many high school curricula do not include long-term inquiry research projects in which students design and implement a lengthy experiment themselves (Leonard and Chandler 2003; Taraban et al. 2006). This could be why postsecondary science instructors find college freshmen to be lacking in basic scientific processing skills (ACT 2009).

Organization of This Handbook

The *STEM Student Research Handbook* was written to support you and your students in two areas of STEM research: planning and conducting research (Chapters 1–6) and doing statistical analysis and communicating the research results (Chapters 7–12). Here is a brief description of the contents of each chapter:

Chapter 1   "Beginning a STEM Research Project" defines research and provides ideas for how to generate and focus ideas for research topics. It also includes a section on safety and ethical issues in STEM research.

Chapter 2   "Research Design" introduces the components of a basic research design and defines key vocabulary terms such as *experimental groups, constants, quantitative data,* and *qualitative data.*

Chapter 3   "Background Research and Note Taking" helps students organize background research. It discusses how to identify reliable resources, provides two different ways to organize notes, offers tips for avoiding plagiarism, and cites helpful technology tools.

Chapter 4   "Writing Hypotheses" walks the student through the process of constructing a testable hypothesis.

Chapter 5   "Proposal Writing" guides students through the process of refining their research designs to a point where they can present you with a written proposal for their experiments that you find acceptable. This chapter also addresses some common misconceptions about scientific writing.

Chapter 6   "Organizing a Laboratory Notebook" introduces students to the essential contents of a laboratory notebook and provides tips about making the same kind of entries in their notebooks that STEM professionals make.

Chapter 7   "Descriptive Statistics" gives instructions for how to find measures of central tendency and statistical variability. It also presents additional calculations that can be performed on data once an experiment is complete.

Chapter 8   "Graphical Representations" describes various graphical representations for communicating data such as various types of graphs, plots, and tables. Graphical representations are organized by the type of data, either qualitative or quantitative.

Chapter 9   "Inferential Statistics and Data interpretation" introduces various mathematical tests that can be used to determine the statistical signifigance of data. The last part of this chapter helps students interpret their data.

Chapter 10   "Documentation and Research Paper Setup" introduces the basic concepts behind documentation and provides a brief introduction to MLA style documentation.

Chapter 11   "Writing the STEM Research Paper" walks students through the writing of the parts of a scientific research paper.

Chapter 12   "Presenting the STEM Research Project" supports students who are asked to make oral presentations about their research.

Organization of Individual Chapters

Each chapter begins with a list of Key Terms that will be introduced in that chapter. The Learning Objectives at the beginning of the chapter will help you determine what it is that your students should know after reading and working through the chapter. Each chapter concludes with Chapter Questions and Chapter Applications. The Chapter Questions align with the Learning Objectives; you might assign them for students to complete as homework or use them to stimulate a class discussion. The Chapter Applications summarize how the ideas taught in the chapter apply to the student research project, including how they apply to the next step of the project.

To the High School Student

I wrote this research handbook for students like you who are being asked to conduct a large-scale research project. In your class, the research project may be part, or even all, of your grade. It is possible that this project is the biggest, and the most long-term, assignment you have ever been asked to complete. Take heart! *STEM Student Research Handbook* is designed to support you in two major research areas: first, it will help you plan and conduct your own experiment (Chapters 1–6) and, second, it will help you analyze and then communicate your results (Chapters 7–12).

Your teacher may choose to introduce you to STEM research in a variety of ways. This handbook addresses many of them; however, be sure to take the advice of your teacher and follow any additional guidelines given to you regarding this research project. Some of you will be asked to pair up with a mentor—either a university researcher or an individual in industry—in addition to having the support of your teacher. Therefore, there will be times in this handbook when I use phrases such as, "Your teacher…." If you are working with an out-of-school mentor, however, substitute the word *mentor* for *teacher*. Ask your teacher whom you should report to for specific aspects of the project.

The fields of science, engineering, mathematics, and technology (STEM) require individuals who are creative and flexible and have a sense of humor about themselves and the world around them. The best way to learn about this exciting side of research is to experience it yourself rather than having it explained to you. I hope that as you begin the journey of developing your own research project, you will see how applicable and exciting research can be and how absolutely critical it is for researchers (including you!) to be creative as well as rational.

Group and Technology Icons

Group icon

Throughout the *STEM Student Research Handbook,* you will find two important icons. The first is the group icon that highlights tips geared to students who are working in pairs or small groups. Overall, the handbook is written with the assumption that most of you will be working individually. Sometimes, however, teachers decide to have groups of students work together on research projects. The group icon is placed next to suggestions regarding how groups can work well together, share research responsibilities, and remain accountable to one another.

Technology icon

The technology icon highlights tips in the text for using technology during the research process. The internet has truly made our world more global, with access not only to mind-boggling amounts of information but also to free Web 2.0 tools to help find and organize research and streamline collaboration among people working together. On this "read-write-web," users can mark up web pages, share files, photos, and create their own interactive spaces. The technology icon brings your attention to specific ways that technology might help you at specific times during the research journey. There are amazing free online

xxvi

tools that can help you retrieve and organize information from the internet and collaborate virtually with others.

You will also find suggestions of specific hardware or software technology that would be helpful during the analysis phase of research. At that point, you will most likely need either a graphing calculator and/or spreadsheet software. Find a way *now* to have access to either one of these technologies so you aren't scrambling later to come up with them. Always communicate with your teacher regarding how much and which technology he or she wants you to use.

Using Technology as Part of the Research Process

Below are introductory descriptions of three basic Web 2.0 tools that will be referred to often in this handbook. If you have a basic understanding of each one, you can make an informed decision about whether or not to use them during the research process.

Wikis

Wikis are websites that allow their members to have web pages that can easily be edited using "what you see is what you get" (WYSIWYG) editing tools. By clicking on the edit button located at the top of each page, a member can type in text, insert images, embed videos, upload documents, and provide links to pages within the wiki or to external web pages. After hitting save, the member has created a web page! If the creator of the wiki wants it to become a place of collaboration, he or she can invite others to become coeditors. Now group members have a place to edit text, add content, and make comments— all on the same web page. *Wikispaces.com* and *pbworks.com* are two sites that are often used for educational purposes.

But that is only the beginning. Using the "history" and "discussion" tabs at the top of each page is what makes a wiki unique. The history tab keeps a detailed record of each edit made to that specific page. The edits are listed by date and indicate the member who made the edit. Edits can be viewed, and added text will be highlighted in one color and anything deleted will be highlighted in another color. By clicking the link titled "revert to this version," edits can be overridden. The discussion tab allows members to discuss the construction of the page. For example, one student might start an outline on the actual wiki page and then also start a discussion "behind" the page, asking other members to add to specific parts.

If you are working on your research project with other students, the wiki can become the unifying place where you are able to meet online as well as the one place where everything is located. You can post schedules and deadlines, discuss protocols, share online sources (although social bookmarking sites do this better), or begin posting drafts of the paper (although Google Docs does this better).

Google Docs

Google Docs is a public place where anyone with an account can upload documents to the web for storage and sharing purposes. A document is loosely defined as any file, including, for example, spreadsheets, images, PowerPoint presentations, and Word documents. Documents can be accessed anywhere a person has internet access, even on a phone. At the very least, it is a place where you should periodically post your proposal or STEM research paper for safe storage during the writing process. Google Docs, like wikis, allows you to invite other people to view or edit the uploaded document. This is a great way to work collaboratively with group members on the same document without worrying if the "version" of the file is the most current one.

Sending documents as attachments via e-mail may be more private, but it is less efficient. Keeping Google Docs organized is easy because files can be put into folders, much as you do on your own computer, and whole Google folders can be shared. If your group decides to make a wiki, you could even organize one of the wiki pages to link to the Google Docs that the group is currently working on. Although Google Docs is the share location that I will refer to throughout this book, you may have access to alternatives, such as a network drive, or a classroom space within a school portal. Websites such as Buzzword, Zoho, Zimbra, and Microsoft Office Live provide similar services.

Social Bookmarking

Social bookmarking—an online bookmarking system—will save you time as you begin researching your topic and developing the research design. If working individually, any bookmarking system (if organized into appropriate folders) is sufficient. However, if you are working with others on this research project, consider a social bookmarking site like Diigo (*www.diigo.com*) or Delicious (*www.delicious.com*). These sites allow you to organize your bookmarks as well as leave comments on web pages, highlight text, and share those bookmarks and edits with the members of your group.

References

ACT. 2009. Focusing on the essentials for college and career readiness: Policy implications of the 2009 ACT National Curriculum Survey results. Retrieved May 16, 2011, from *www.act.org/research/curricsurvey.html*

Ayers, J. M., and K. M. Ayers. 2007. Teaching the scientific method: It's all in the perspective. *American Biology Teacher* 69 (1): 17–21.

Bereiter, C., and M. Scardamalia. 2009. Teaching how science really works. *Education Canada* 49 (1): 14–17.

Drake, K. N., and D. Long. 2009. Rebecca's in the dark: A comparative study of problem-based learning and direct instruction/experiential learning in two 4th-grade classrooms. *Journal of Elementary Science Education* 21 (1): 1–16.

Leonard, W. H., and P. M. Chandler. 2003. Where is the inquiry in biology textbooks? *American Biology Teacher* 65 (7): 485–487.

Marcus, J. M., T. M. Hughes, D. M. McElroy, and R. E. Wyatt. 2010. Engaging first-year undergraduates in hands-on research experiences: The Upper Green River Barcode of Life project. *Journal of College Science Teaching* 39 (3): 39–45.

Tang, X., J. E. Coffey, A. Elby, and D. M. Levin. 2010. The scientific method and scientific inquiry: Tensions in teaching and learning. *Science Education* 94 (1): 29–48.

Taraban, R., C. Box, R. Myers, R. Pollard, and C. Bowen. 2006. Effects of active-learning experiences on achievement, attitudes, and behaviors in high school biology. *Journal of Research in Science Teaching* 44 (7): 960–979.

Tarhan, L., H. Ayar-Kayali, R. O. Urek, and B. Acar. 2008. Problem-based learning in 9th grade chemistry class: Intermolecular forces. *Research Science Education* 38: 258–300.

Wong, K. K., and J. R. Day. 2009. A comparative study of problem-based and lecture-based learning in junior secondary school science. *Research in Science Education* 39 (5): 625–642.

Beginning a
STEM Research Project

Introduction

Science, technology, engineering, and mathematics (STEM) research refers to experiments conducted to address problems in those fields that can be tested using the scientific method. The *scientific method* is an inquiry process used to systematically study, investigate, and to provide explanations for observed phenomenon in the natural world. This method is used by STEM professionals to answer questions they have about important world problems and usually includes carefully orchestrating a situation that allows them to observe, measure, and test their ideas (Valiela 2001). This "situation" is known as an *experiment*. Most experiments include a hypothesis; a variable that can be manipulated by the researcher; and variables that can be observed, measured, calculated, and compared. When possible, these experiments are completed in a controlled environment.

> ## Learning Objectives
>
> By the end of this chapter, you should be able to
>
> 1. identify resources that can be used to generate research ideas,
>
> 2. list possible research topics,
>
> 3. put preliminary research ideas into testable questions, and
>
> 4. apply safety and ethical issues to your own project ideas.
>
> *Note to the teacher:* These objectives are restated in the form of questions at the end of each chapter (e.g., see p. 11). The questions can be used to check for understanding after the class has completed the chapter.

Key Terms

Data: The measurements and observations that are collected as part of a research project, often a combination of measurements and descriptions.

Dependent variable: A dependent variable is what you measure in the experiment and what is affected during the experiment. The dependent variable responds to the independent variable. It is called dependent because it "depends" on the independent variable. In a scientific experiment, you cannot have a dependent variable without an independent variable. (Source: *www.ncsu.edu/labwrite/po/dependentvar.htm*)

Entity: The subject, specimen, or item that is studied as part of a STEM research project.

Experiment: The test conducted as a part of the scientific method that includes a hypothesis; a variable that can be manipulated by the researcher (independent variable); and variables that can be observed, measured, calculated, and compared (dependent variables).

Independent variable: The variable you have control over, what you can choose and manipulate. It is usually what you think will affect the dependent variable. In some cases, you may not be able to manipulate the independent variable. It may be something that is already there and is fixed or something you would like to evaluate with respect to how it affects something else. Example: You are interested in how stress affects heart rate in humans. Your independent variable would be the stress and the dependent variable would be the heart rate. You can directly manipulate stress levels in your human subjects and measure how those stress levels change heart rate. (Source: *www.ncsu.edu/labwrite/po/independentvar.htm*)

Scientific method: The scientific method is an inquiry process used to systematically study, investigate, and provide explanations for observed phenomenon in the natural world.

STEM research: Experiments that test hypotheses in science, technology, engineering, and/or mathematical fields.

The word *data* refers to the measurements and observations that are collected as part of a research project. The kinds of measurement data commonly collected in STEM fields are *acidity/alkalinity, area, circumference, density, electrical current/potential/resistance, force, growth (time, weight, volume, length/width), heat, humidity, light intensity, mass, pressure, sound intensity, temperature, time, velocity, volume, and weight.*

Data can also be collected by describing observations and using words and photographs. In that case, one asks, How does [something] look, smell, sound, feel, and taste (when appropriate)? These types of observations supplement the measurements taken throughout the experiment. A combination of measurements and descriptions are listed to determine whether the proposed idea of the experiment is supported by the data.

You have probably taken many science courses where the first chapter of your textbook presents the scientific method. Have you ever noticed that the "steps" in the scientific methods are rarely ever worded in the exact same way? Why would that be? Isn't science supposed to be about accuracy and step-by-step procedures? First you do this, then you do that, so that you can collect this and write about that. So here's my challenge to you: Do not think about the scientific method as steps or procedures that a scientist must complete before he or she gets answers to the questions that were asked.

Instead, think of the scientific method as doing your absolute best to answer a question with the knowledge, skills, resources, and technology you have available. Could someone else, with different skills and technology, answer the same question in a different way? Absolutely. But this should not keep you from answering it your way. Could it mean that your design will be obsolete in a couple of years when new technology is available? You bet! But this is no excuse for not doing what you can with what you have. This also highlights the importance of STEM professionals sharing what they have learned, both successes and failures, through journal articles and conference presentations, so that everyone doing research in that area can benefit from what each has learned.

Real research can be messy. Therefore, when I talk about the scientific method, or the general research process, I use the word *stages* rather than *steps*. The word *step* suggests each step is an equal distance from each other, and that once you have reached the top step, you should be embarrassed to

Figure 1.1

Stages of STEM Research

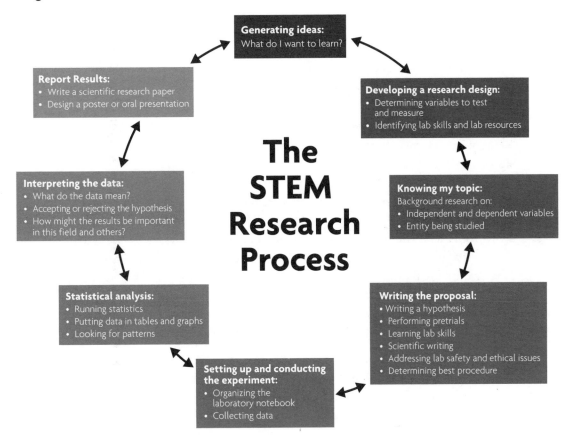

have to go back to the first step. But in real science, this is exactly what happens. Learning, by its very nature, can lead us in many directions, and often, it is in the moments when we have to go back and rethink or retest something we previously studied that the best discoveries can be made.

While moving through the stages of the scientific research process, you will soon learn that in scientific inquiry, the more you know, the more you know you don't know. That is, as you gain more insight to a problem, you usually come out with more questions than you do answers. It is even possible that you do not answer the question you started out to answer but another question entirely. And for most STEM professionals, the emergence of new questions is celebrated, not frowned on. However, take heart. It is likely that your teacher, who is overseeing your research project, wants to help you develop a research study that is not too messy. So, as you get feedback from your teacher about possible modifications to your research design (see Chapter 2), it probably is in your best interest to consider them.

Generating Research Ideas

More than likely, you are curious about various things, including things that you have seen recently. Maybe it was something you saw in a movie, in the news, or around your house or school. You wondered to yourself, *I wonder if that is really possible* or *I wonder if the same thing would have happened if….* Questions like these are the seeds of great research ideas. Thinking about how the world functions, and how you might improve it, is at the heart of STEM research. The first stage of generating a research idea is to determine several entities you might be interested in studying. For the purposes of this handbook, the term *entity* refers to the subject, specimen, or item that you will study for your STEM research project.

Getting started on a research project often brings with it two different dilemmas: Either students have no idea what they want to study or they have a very specific idea of what it is that interests them. I'd like to warn you to stay away from each extreme. It is best to have a general idea of what you want to study—that way you can focus your research—but you will not be so narrow in your thinking that you miss a great research opportunity.

This book doesn't include a list of specific research ideas from which to choose. There are plenty of other resources that contain research topics. However, here are some general tips on how to generate a research topic:

- Choose a topic that is interesting to you. Maybe there is a topic that you have always wanted to know more about. You will be working with this topic for a long time, so choose it carefully.

- Search for ideas on the internet. Look not only for research projects that have already been conducted but also for general information about the entity you might study and the items you might manipulate within the experiment.

- Reflect on a topic you heard about on television that piqued your interest.

- Think about issues your family deals with. Maybe there are personal reasons why you might be interested in a specific topic.

- Flip through a science or math book, magazines such as *Science News*, or encyclopedias for inspiration.

- Borrow your teacher's science supply catalogs (such as those from companies like Carolina Biological Supply or Flinn Scientific) and look through the different live specimens, chemicals, and apparatus that you could purchase or borrow.

- Think about the lab skills that you have already learned. How might you use those skills in a research study? Are there other skills that could be easily learned that you'd like to try?

- Ask your teacher for a list, or maybe a tour, of available equipment in your school that you could check out to use at home or use within the school's lab.

Many times just knowing the tools, instruments, or tests that are available to study certain topics can spark an idea for a research project. Use the list in Table 1.1 (pp. 6–7) to consider entities within STEM fields that you might study. Though you may think you have limited access to some of the equipment listed here, do not underestimate your ability to improvise. You may be able to design an instrument to measure what you want. And there are technologies you already have that you might use, such as your graphing calculator or cell phone. A calculator-based laboratory (CBL) system or calculator-based ranger (CBR), along with probes, might be easy to obtain from various departments at your school. Smart phones with inexpensive applications (apps) may also help you measure something if you do not have access to more expensive equipment.

Once you have a general topic, start asking yourself questions. Let your natural curiosity lead you to possible ideas to study. However, stay away from "why" questions—for example, "Why do more algae seem to grow in slower moving stream water?" They tend to be too broad and worded in such a way that they are not testable. Instead, you can rephrase a question to break it into

Table 1.1

Sample STEM Topics With Associated Tools, Instruments, and Tests

What to Measure	Tool, Test, or Instrument
Absorbance	Spectrophotometer (Spec-20)
Acidity/alkalinity	pH paper, pH meter
Altitude	Altimeter
Angles of slope/tilt	Clinometers, protractor, sextant, transit, goniometer, Geometer's Sketchpad
Area	Meterstick (with appropriate formulas), planimeter
Bacteria	Gram stain, incubator, hemocytometer, spectrophotometer
Blood pressure	Sphygmomanometer
Calculus	Calculus modeling and equation solving systems (Mathematica), graphing calculator
Color/pigments	Chromatography
Conductivity	Amperometer, potentiometer
Density	Balance and meterstick, pycnometer, hydrometer
Earth movements	Seismometer
Electrical current	Ammeter, multimeter, galvanometer
Electrical potential	Voltmeter, multimeter, galvanometer
Electrical resistance	Ohmmeter, wheatstone bridge
Embryology	Chick incubator
Force	Spring scale, dynamometer
Global position	Global positioning system (GPS)
Heat	Calorimeter
Humidity	Hygrometer
Insects (trapping)	Berlese funnel, bait trap, aspirator, sand/mud sieve, nets
Length/width	Meterstick, tape measure, micrometer, Vernier caliper
Light	Spectrometer, photometer, light meter, photoelectric cell
Mapping	Transit
Mass	Spring balance, lever-arm balance, electric balance
Mathematics	Geometer's Sketchpad, GeoGebra, statistical software, Mathematica, graphing calculator, TinkerPlots, Fathom

Table 1.1 *(continued)*

Sample STEM Topics With Associated Tools, Instruments, and Tests

What to Measure	Tool, Test, or Instrument
Muscle activity	Electromyography, video analysis
Muscle tone	Myotonometer
Optical density	Photoelectric colorimeter
Photosynthesis (rate of)	pH meter, chromatography
Plant growth	Time-lapse camera, metric ruler, protractor
Pressure	Barometer, manometer, mechanical pressure gauge
Radioactivity	Geiger counter, imization detector
Range of motion	Goniometer
Respiration	Respirometer
Salinity	Salinometer
Small live specimens	Light microscope, stereomicroscope, magnifying glass
Soil	Soil coring tube, screen sieve, soil thermometer, chemical tests (phosphate, nitrogen, potassium, pH)
Soil porosity	Soil samples, beakers, water
Sound intensity	Audiometer, decibel meter, sound-level meter
Statistical comparison	Statistical software like Excel or SPSS, Fathom
Temperature	Thermometer, infrared thermometer, thermocouple, thermistor, pyrometer
Tensile strength	Spring scale, metric ruler
Time	Stopwatch, timer, watch
Transpiration	Graduated cylinder, closed container
Tree diameter	Diameter tree measuring tape
Tree height (or other large entity)	Tangent height gauge
Tree wood quality/growth rate	Increment borers
Turbidity	Secchi disk, turbidity tube, turbidity meter
Velocity	Speedometer, anemometer, stopwatch/meterstick, stream flow meter
Viscosity	Stopwatch, calibrated tube, ball, funnel
Volume	Graduated cylinder, pipette, burette, volumeter, manometer

(continued)

Table 1.1 *(continued)*

Sample STEM Topics With Associated Tools, Instruments, and Tests

What to Measure	Tool, Test, or Instrument
Water collecting debris	Manta trawl
Water flow	Water flow probe or ball, meterstick, stopwatch
Water quality	Dissolved oxygen (kit or titration), pH
Water retrieval @ depths	Water sampler
Water turbidity	Secchi disc
Weight	spring scale, electric scale, balance
Wind speed	Anemometer

smaller parts, which *are* scientifically testable—e.g., "Which stream velocities encourage more growth of algae?" That question is now a measureable and, therefore, testable question. *Note:* Testable questions often begin with *How, What, When, Who,* or *Which.* Write several questions that you might be interested in studying. The Southwest Center for Education and the Natural Environment has an inquiry tutorial that can help you write some preliminary research questions *(http://scene.asu.edu/habitat/inquiry.html).*

Focusing Preliminary Research Topics

Once you have a preliminary research topic, you will need to focus it, using a new group of questions. Finding answers to each of these questions will help you get closer to what it is you will eventually study. Figure 1.2 shows the connection of the following questions: What entity should I study? What could I manipulate or change? What effects could I measure? What skills, knowledge, and tools would I need?

It is likely that your teacher will encourage you to narrow your "effects" (things to count and/or to describe) to only one or two items, most likely one or two you will measure and one or two you will describe. Use Table 1.2 as a guide as you complete "Student Handout #1: Focusing Preliminary Research Ideas," page 14. The goal of this stage is to brainstorm various combinations of the same research topic. Try and list as many answers to the questions in Figure 1.2 as possible. Don't worry yet about what will work, just generate ideas.

You may want to fill out Student Handout #1 several times, comparing a couple of ideas you are considering at this time. Your teacher may have you

Figure 1.2

Focusing a Research Topic

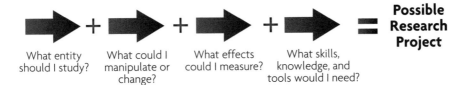

What entity should I study? + What could I manipulate or change? + What effects could I measure? + What skills, knowledge, and tools would I need? = **Possible Research Project**

enter these preliminary ideas into your laboratory notebook (see Chapter 6). In addition to the various issues raised earlier in this chapter, you also need to pick a topic that is both safe and ethically responsible, the subject of the following section.

Safety and Ethical Issues

Safety Issues

Here at the beginning of the research process, you need to carefully consider the safety and ethical issues involved in conducting a STEM research project. Special safety considerations must be made when working with chemicals, mold and microorganisms, electricity, radiation, and vertebrate animals—including humans. Follow the safety precautions you have learned from your experience in completing laboratory research in your teacher's classroom, but also read up on additional information on safety. You must understand not only a particular safety rule, but also why it is important.

Ethical Issues

In addition to lab safety, you must also consider the ethics of conducting research. Research ethics involve understanding the "norms of conduct that distinguish between acceptable and unacceptable behavior" (Resnik 2010, p. 1). Ethics in the context of conducting research might be something you have never had to deal with on a personal level because the lab experiments your teachers have had you do were chosen so they did not violate any ethical issues.

Table 1.2

Sample Preliminary Research Ideas on the Topic "Seeds"

Entity to Study: Seeds
What I could manipulate or change: ☐ amount of sunlight ☐ quality of sunlight (color or opacity) ☐ size or material of container ☐ temperature and humidity ☐ amount of water
What effects I could look for (things to count as well as describe): ☐ seedling growth ☐ speed of germination ☐ # of new leaves ☐ # of leaves per stem ☐ root length ☐ speed of root growth ☐ color of seed, root, stem, and leaves ☐ health of seedlings
Knowledge, tools, and skills I would need to do the project: ☐ How does a seed germinate? What do seeds need to germinate in a controlled environment? ☐ What types of seeds would be good to use for a germination study? ☐ How do I study germination without affecting the root and stem structures? Should I plant the seeds in soil or something else?

At universities, however, before researchers conduct experiments, they must receive training and then file certain papers (documentation) with a local Internal Review Board (IRB). University researchers planning to do research with vertebrate animals must receive training and then file documentation with an Institutional Animal Care and Use Committee (IACUC) to be sure that animals will be treated humanely.

The main concern about beginning researchers (such as yourself) is that they are not aware of the issues that may violate the "acceptable behavior" norm. Therefore, there are national regulations put in place to ensure that researchers have thoroughly thought through their experimental procedures, understand the safety and ethical issues, can justify their methods, and can ensure the humane treatment of the entities being studied. Local, state, national, or international fairs and symposia[*] interpret these regulations and provide guidelines for their research participants. Your teacher will provide resources to help you obtain any documentation and committee approval that your experimental design may require for competition.

Even if your project is not being submitted for a competition, safety and ethics must still be considered. Since the Intel International Science and Engineering Fair (ISEF) (*www.societyforscience.org/isef/document*) and the Junior Science and Humanities Symposium (JSHS) (*www.science.siu.edu/ijshs/pdf/ijshs.pdf)* are the largest organizations to host high school state fairs, the guidelines and documentation posted on their websites are excellent models to follow.

If you are working with a mentor in industry or at a university, follow the safety and ethical guidelines set forth by that organization.

Human Subjects as Research Entities

While studying human subjects may be intriguing, procedures must be taken into consideration that protect the rights and welfare of the participants. Most human subject studies require informed consent or assent from the research subjects and IRB approval. Informed consent or assent[**] is the process by

> ### Resources for Laboratory Safety
>
> - Hurr, A. K. 2000. *CRC Handbook of laboratory safety.* 5th ed. Boca Raton, FL: CRC Press.
>
> - Flinn Scientific: *www.flinnsci.com/Sections/ Safety generalLaboratorySafety.asp*
>
> - Princeton University: *web.princeton.edu/sites/ ehs/labsafetymanual/index.html*
>
> - Duke University & Duke Medicine: *www.safety.duke.edu/safetymanuals/Lab/ default.htm*
>
> - World Health Organization: *www.who.int/csr/resources/publications/ biosafety/WHO_CDS_CSR_LYO_2004_11/en*

----

[*] Symposia are formal meetings at which experts discuss a particular topic.

[**] Both *consent* and *assent* basically mean "agreement," but *consent* is used more often in regard to legal matters (e.g., "the age of consent"). *Assent* generally means a positive and voluntary agreement (e.g., "I gave my assent to the plan").

which researchers inform potential study participants about a study and gain verbal or written consent from those participants. If you are asking an individual under the age of 18 to participate, you must also obtain consent from the parent or guardian of that individual.

As you begin writing your proposal in Chapter 5, remember to include IRB committee approval information within the methods section and provide polished versions of consent or assent letters in an appendix to the proposal. Although you need to check regulations in your local area or state, some human subject studies *may not* require IRB approval. These studies may include the following:

- Testing of a student-designed invention, where participant feedback is not personal data and does not pose a health risk.

- Data or record review studies, where preexisting, publically available data sets are used that do not involve interaction with human subjects.

- Public behavioral observations of individuals (e.g., shopping mall, public park) in which all the following conditions apply. The researcher

    o has no interaction with the individuals being observed,

    o does not manipulate the environment in any way, and

    o does not record any personally identifiable data

Other Restricted-Research Entities

Nonhuman vertebrates, human subjects, and potentially hazardous biological agents have specific national regulations that must be followed. Organizers of various science fairs or symposia accept these topics in different ways. Some do not allow high school students to study within these topics at all; others place specific restrictions and require certain documentation. Table 1.3 (p. 12) lists subcategories within a restricted research topic, requirements that may be made by a fair or symposia organization, and alternative project ideas.

Chapter Questions

1. List at least three places you can go to get ideas for your research project.

2. Describe your top three research topic ideas.

3. Focus these three preliminary research ideas into testable questions.

4. How do the safety and ethical issues discussed in this chapter apply to your project idea?

Chapter Applications

Once you have read this chapter, I hope your curiosity will lead you to several research topics. You should have completed Student Handout #1 (p. 14) at least once, if not several times. After completing the handout, you might brainstorm with your classmates and family to improve on the ideas you already have. On the completed handout, begin to jot down various questions using different variations of what you could manipulate and what effects you

Table 1.3

Restrictions, Possible Requirements, and Alternatives for Certain Research Topics

Restricted Research Topic	Possible Requirements	Alternative Topics
Nonhuman Vertebrates • Mammalian embryos or fetuses • Tadpoles • Bird and reptile eggs • Fish • Mammals	• Supervision by a qualified scientist or designated supervisor. • Veterinarian consultation during experiment. • Approval by the Scientific Review Committee (SRC) and/or by the International Animal Care and Use Committee (IACUC) before research begins.	• Do similar studies on other organisms such as algae, ants, beetles, crabs, crayfish, protists, fruit flies, houseflies, lichen, yeast, vinegar eels, slugs, snails, earthworms, planaria, or mealworms.
Human Subjects • Physical activities • Psychological, educational, and opinion studies (e.g., surveys, questionnaires, tests) • Behavioral observations that include interaction, are in nonpublic locations, collect identifying data (e.g., name, date of birth)	• Review and preapproval by the Institutional Review Board (IRB). • If medical information is involved, compliance with HIPAA (health information privacy) regulations. • Risk assessment. • Consent/assent forms for participants to complete. • Supervision by a qualified scientist or designated supervisor.	• Obtain data that are already publically available. • Develop a research design that does not include physical activity or a design that uses data that are anonymous (not able to be linked to a particular participant)
Potentially Hazardous Biological Agents • Microorganisms (bacteria, viruses, viroids, prions, fungi, and parasites) • Recombinant DNA (rDNA) • Human or animal (fresh or frozen) tissues • Blood • Body fluids	• Risk assessment. • SRC and/or IACUC approval before research begins. • Human and Vertebrate Animal Tissue Form (at *www.societyforscience.org.isef*).	• Obtain tissue cultures from reputable biological supply houses. • Use baker's yeast, bacteria, or fungi that are approved by the International Science and Engineering Fair (ISEF) organization. • Use plant tissue, meat or meat products, hair, fossilized tissue or archeological specimens, or prepared fixed tissue.

Source: Adapted from International rules for precollege science research: Guidelines for science and engineering fairs. 2010. *http://apps.societyforscience.org/isef/rules11.pdf*

could look for. Remember, the questions need to be narrowed into scientifically testable questions. For example, if you were investigating a topic about microwave emissions from cell phones, you would ask, "How do microwave emission levels vary for new cell phones compared to older ones?" or "What type of cell phone use—calls or texting—emit more microwaves?"

Later, you will modify these questions into a formal hypothesis. Now, however, is the time to explore research topics and possible variations of an experiment. In the next chapter, you will learn the specifics of designing a research experiment. Although you may not have your research topic finalized yet, once you begin developing a research design, you will need an official research topic.

References

Gordon, J. C. 2007. *Planning research: A concise guide for the environmental and natural resource sciences.* New Haven, CT: Yale University Press.

The inquiry process. 2004. Retrieved March 30, 2011, from *http://scene.asu.edu/habitat/inquiry.html.*

International rules for precollege science research: Guidelines for science and engineering fairs. 2010. Retrieved March 16, 2011, from Society for Science and the Public, Intel ISEF document library website: *http://apps.societyforscience.org/isef/rules/rules11.pdf.*

Pedroni, J. A., and K. D. Pimple. 2001. A brief introduction to informed consent in research with human subjects. Retrieved March 17, 2011, from *http://poynter.indiana.edu/sas/res/ic.pdf.*

Resnick, D. B. 2010. What is ethics in research and why is it important? Retrieved March 15, 2011, from *www.niehs.nih.gov/research/resources/bioethics/whatis.cfm.*

Valiela, I. 2001. *Doing science: Design, analysis, and communication of scientific research.* New York: Oxford University Press.

Focusing Preliminary Research Ideas

Name _____ Class _____ Date _____

Directions: Answer the following questions based on your research interest right now.

1. What entity do I want to study?

2. What could I manipulate or change?

3. What "effects"—that is, things to count as well as to describe—could I look for?

4. What knowledge, tools, and skills would I need to do this project?

NATIONAL SCIENCE TEACHERS ASSOCIATION

—2—

Research Design

Introduction

In Chapter 1, you worked on focusing your preliminary research ideas. In this chapter, you will learn how experiments are structured. This structure or experimental setup is called the *research design*. The research design of an experiment determines both whether the experiment is likely to succeed and the reliability of its results.

Learning Objectives

The main objective of this chapter is to have you write a first draft of your experimental research design. By the end of the chapter, you should be able to

1. list the main components of an experimental design,

2. describe the purpose of having a hypothesis in a STEM-based research project,

3. explain the importance of doing background research on independent and dependent variables,

4. compare and contrast the individual entities or trials within the experimental groups,

5. describe how constants are different from the control,

6. explain why it is important to consider possible extraneous variables when you are designing a STEM research project,

7. distinguish between quantitative data and qualitative data, and

8. describe how recording only inferences may interfere with data collection.

Key Terms

Constants: The factors within an experiment that are kept the same for all groups or trials in an attempt to reduce the influence of extraneous variables.

Control group: The group in an experiment that receives the exact treatment as the experimental groups *except* it does not receive any change of the independent variable. It is the group to which the experimental groups are compared.

Dependent variable (DV): The variable in an experiment that changes *in response to* the independent variable and, therefore, is also referred to as the *responding variable.*

Experimental groups: The groups or trials in an experiment that receive all the same conditions *except* varying amounts or qualities of the independent variable.

Extraneous variable: An "undesirable" variable in addition to the independent variable that may influence the results of an experiment, introducing error if it is not, as much as possible, controlled or significantly decreased in the research design.

Focal sampling: A behavioral recording technique where a *narrative* (i.e., what is called an *essay* in English class) is written on every behavior of one individual or group for a set length of time.

Hypothesis: A tentative (i.e., not final and definite) and testable proposed explanation for an observable phenomenon.

Independent variable (IV): The variable in an experiment that is purposely changed or manipulated, either in quantity or quality, by the researcher; also referred to as the *manipulated variable.*

Inference: A conclusion, based on facts, that a person perceives to be true.

Population: The complete collection of every item that has the same characteristics of the individuals in the sample group.

Qualitative data: Data that describe characteristics or qualities, such as color, odor, or texture, or data that describe category frequency or ratings, such as stem sturdiness (e.g., "sturdy," "somewhat sturdy," "limp").

Quantitative data: Data that use numbers with a unit of measurement, such as the length of an insect in millimeters (millimeter is the unit of measurement) or the weight of a projectile in kilograms (kilograms is the unit of measurement).

Sample: A subcollection of data that represent a larger population.

Scan sampling: A behavioral recording technique where the activity of the individual or group is recorded only at preselected time intervals.

Sequence sampling: A behavioral recording technique where behaviors that occur within a sequence are recorded in the order in which they occur.

Trial: The replication of experimental and control groups; used to decrease the influence of variations associated with the independent variable, researcher measurement error, and difference between entities studied.

Components of a STEM Experimental Research Design

An experimental research design includes a hypothesis, variables, experimental and control groups, and constants. Each of these elements is briefly discussed below.

Hypothesis

Once you have determined the question you would like to answer, and after you have begun background research, you are ready to modify your question into a testable statement. You do this by writing a *hypothesis*, which is a tentative, yet testable, proposed explanation for an observable phenomenon or event. The purpose of the hypothesis is to formulate what you want to test and defines the limit of your experiment. It is considered tentative because it states a connection that you believe exists and want to test. However, one research experiment will not ultimately "prove" or "disprove" the connection you are suggesting. The purpose of a hypothesis is to *connect the manipulated changes made by the independent variable with the effects on the measurements of the dependent variable.*

Writing hypotheses to be tested through experiments and observations is central to doing research (Gordon 2007). The question you developed in Chapter 1 will help you stay focused as you do your background research. Now, by changing the question into a hypothesis statement, you accomplish several critical research design issues. In writing a hypothesis, you will

1. determine a specific variable to be tested,

2. determine how changes within the experiment will be measured or recorded, and

3. predict an outcome of what you think the results of the experiment will be.

For instance, in the following planaria example, the question only asks "how" reproduction is affected by temperature and is not written in such a way that it could be answered in a single experiment. How the reproduction "effect" would be measured is not clear. Planaria reproduction could be measured many different ways, such as the number of offspring that come from one individual or the mortality rate of offspring. (The independent variable is underlined once and the dependent variable is underlined twice.)

> **Question:** *What effect does temperature have on planaria reproduction?*

Hypothesis: *If the <u>speed of planaria reproduction</u> is related to <u>temperature</u>, then planaria in lower temperatures will reproduce more slowly than planaria in higher temperatures.*

The hypothesis, on the other hand, identifies not only the specific variable to be tested (temperature) but also what will be measured (speed of planaria reproduction). The inclusion of a prediction—that lower temperatures will lower reproduction rates—makes it clear that the experiment is designed to either support or reject that prediction. Hypotheses like this one are testable because (a) one variable is tested, (b) it is clear how the changes will be measured, and (c) it includes a prediction that will be either supported or rejected by conducting the experiment.

You should write your hypothesis after you do your preliminary research but before you begin your experiment. More details about how to write a hypothesis are provided in Chapter 4, "Writing Hypotheses."

Variables

The *independent variable* (IV) is the variable that is purposely changed or manipulated. Information about the independent variable is known before the experiment begins. Independent variables are also known as *manipulated variables* because you change either the quantity or quality of this variable in the experimental groups. Therefore, the independent variable will determine the organization of levels for the experimental groups. Although more complex experiments can have more than one independent variable, research projects that are to be completed over a series of weeks or months should only have one variable that can easily be tested and measured.

The *dependent variable* (DV) is the variable that changes in response to the independent variable and, therefore, is also referred to as the *responding variable*. Essentially, this is the "effect" and the data that you record during the experiment. It is best to quantify measurements as much as possible, but accurate descriptive data throughout the experiment are helpful as well.

Every research question has several ways in which changes could be measured. Background research should help you determine which dependent variables are most likely to show change in the time you have to conduct the experiment. Therefore, it is important to base the choice of your dependent variable on what you have learned in your background research. Plan your experiment so it focuses on a few related dependent variables. For example, for the research question "How effective are plant-based insect repellents?" there are several different options of dependent variables that a researcher could choose.

| Research Question: "How effective are plant-based insect repellents?" ||
Independent Variable	Possible Dependent Variables
Different brands of plant-based repellents (ideally with differing levels of the active ingredient)	• Total number of insect bites • Size of insect bites • Color and/or itchiness of insect bites • Length of time

Experimental Groups

Experimental groups are the treatment groups or trials that receive all of the same conditions, *except* varying amounts or qualities of the independent variable. Experimental groups are sometimes called *treatment groups* because they receive the change of the independent variable. An important component of designing strong experiments is replication (i.e., performing an experiment more than once). In some STEM experiments, experimental groups containing several entities can be running at the same time, while other experiments will have multiple trials, or runs, that are conducted periodically over time.

The word *trials* refers to the number of treatment replications that you perform on experimental and control groups. Having multiple entities in each experimental group or running multiple trials is important because it decreases the influence of variations associated with the independent variable, researcher measurement error, and difference between the entities studied. For example, in a biology experiment with seeds and pH, four experimental groups, each with multiple seeds, can be set up at the same time, each with different pH levels, and data can be collected from each of the groups at regular intervals throughout the experiment. But an engineering research project testing the mechanical advantage of differing arm lengths of a catapult would use multiple trials of each arm length, which would require making adjustments to the catapult in between experimental groups.

By doing thorough background research, you should be able to determine both how many experimental groups to have and the appropriate levels of the independent variable for each of the groups. The organization of the experimental groups is critically important for a strong research design. Having experimental groups that are not varied enough in quantity or quality may not show any change in the dependent variable and, therefore, will not help you determine any connection between your independent variable and dependent variable. Let's look at an example. The following hypothesis is testing to determine a relationship between levels of vitamin C and when a fruit is picked.

If the concentration of vitamin C in oranges is related to the length of time it has been removed from the tree, then oranges freshly picked will have higher levels of the vitamin.

The *experimental groups* for this experiment should be *varying times after the fruit is picked.* It is important that these groups be selected carefully to show an adequate spread of results. If the vitamin C levels are measured in oranges at intervals of 6 hours, 12 hours, and 24 hours, the levels of vitamin might not differ enough to notice. Similarly, experimental groups divided into extreme high and low quantities will not show the detail needed to analyze the effects. If experimental groups in the orange/vitamin C experiment are measured at 1 day, 4 weeks, 8 weeks, and 16 weeks, the levels of vitamin C might be drastically different, but, without multiple gradients of the independent variable, there is no subtle data to determine critical levels or provide insight as to why those changes might have occurred. Therefore, it is important that you use background research to study the variables so that your groups can be set up appropriately.

It is also possible that experimental groups cannot be predetermined. Sometimes, it is only after the data are collected that data can be grouped for analysis. For the river otter experiment shown in Figure 2.1, the data could be categorized into groups based on the range of temperatures that were actually observed on data collection days. The independent variable of this experiment is the change in water temperature and the dependent variable is the frequency of river-otter behaviors that the researcher categorizes as active or nonactive when making observations.

Figure 2.1

Sample Behavioral Ecological Design Table

Hypothesis Draft				
Behavior of a river otter in different water temperatures				

Independent Variable	Background Questions			
Water and air temperature	What is the normal river otter behavior? How should I record and quantify behavior in my lab notebook? How quickly does water temperature change in the fall? (Will it be significant enough to study?) How do otters prepare for the winter months? Behaviorally and physically?			

Dependent Variable	Constants			
Quantitative None *Qualitative* Descriptions of behaviors Tallies of specific behaviors (categorical data)	Make observations at the same time of day Same otter observed in the same location throughout the experiment Note: Experimental and control groups can't be predetermined! I will categorize after data are collected, choosing temperature category ranges that match the observation data.			

Experimental Groups and Control Group	Control Group	Exp. Group #1	Exp. Group #2	Exp. Group #3
	Average expected temperature for the season			Coolest water temperature

Control Group

The *control group* is the *one* group to which all other groups will be compared. The control group receives the exact treatment as the experimental groups *except* it does not receive the change of the independent variable. In the orange/vitamin C experiment, the control group would be the level of vitamin C while still attached to the tree (or shortly thereafter), but in all other ways, it is treated and cared for in the same manner as the experimental groups. This way you can determine whether or not there are hidden variables that may be changing without you knowing it.

A control can also be a known measurement or level of the independent variable. In the river otter experiment, the average expected temperature could be designated as the control group.

Sometimes a control can only be designated after data are collected. Also, for some experiments, there is no control group and the comparison among the experimental groups is enough.

Notice there is no control group for the geology research design shown in Figure 2.2. This is acceptable because the data collected at each depth will be compared to one another because the researcher is looking for a pattern in an event that occurred many years ago. This geology research design also highlights how, because the data are located in the environment, there are fewer constants than there would be if completed in a controlled setting.

In the chemistry design shown in Figure 2.3, it is important to note that several trials of this experiment should be performed, and additional experimental group rows could be added in the table to indicate this. Comparing this data to another brand of cosmetics would be a way to expand this experiment.

Both the experimental and control groups are considered a smaller sample of the larger population. Statistically, a *sample* is a subcollection that represents the entire population. The sample is the group from which you actually collect data. The *population* represents a complete collection of every item that has the same characteristics as the individuals in the sample group. For example, in an experiment that has three experimental groups and a control group, these four groups make up a representative sample of the entire population. Understanding that these groups are samples that represent a population is important when you begin to statistically analyze your data after your experiment.

Samples are commonly used in research studies to make claims regarding the entire population. The assumption is that as long as the sample represents the population—that is, that the characteristics of those entities within the sample match those in the population—these types of inferences (claims) can be made. The larger the sample, the more likely this assumption is correct. In the STEM studies we are conducting, rarely if ever, can data be collected from an entire population; therefore data from samples must be studied instead.

Figure 2.2

Sample Geology Research Design Table

Hypothesis Draft				
The number and type of fossils will differ at varying depths.				

Independent Variable	Background Questions			
Soil sample from different depths (meters)	What methods for collecting soil from a cliff will do the least amount of damage? And be safest for me? How will I identify fossils? What categories for "type" will I use? How do I best organize my lab notebook for this type of research? How will I count partial fossils?			

Dependent Variable	Constants			
Quantitative Number of fossils (# per kg) *Qualitative* Type of fossils	Soil samples removed from cliff on the same day Soil samples spread out and allowed to dry before weighing 1 kilogram from each depth			

Experimental Groups and Control Group	Exp. Group #1	Exp. Group #2	Exp. Group #3	Exp. Group #4
	Soil sample from top surface of a cliff	Soil sample at 5 meters	Soil sample at 10 meters	Soil sample at 15 meters

Figure 2.3

Sample Chemistry Research Design Table

Hypothesis Draft			
Comparing color dyes found in cosmetics			

Independent Variable Cosmetics that contain straight dyes	**Background Questions** In what cosmetics are dyes found? Which will be easiest to test? What are the different types of dyes? What testing has been done on dyes? Are some dangerous? How are cosmetics tested? What makes cosmetic dyes stay where they are applied? Can I use chromatography to distinguish the different dyes? How do I do that? What supplies will I need? Are dyes required to be listed in the ingredients? What makes a cosmetic product darker? Different darker dyes or many dyes used together?		
Dependent Variable *Quantitative* # of dyes per product *Qualitative* Description (and photographs) of chromatograms	**Constants** Chromatograms are all made from filter paper used from the same package. Same method of obtaining samples is used for all groups. Same product and brand is used, but different shades and multiple trials. Note: Experimental groups will be various shades from lightest to darkest.		
Experimental Groups and Control Group	**Exp. Group #1** Lightest color product	**Exp. Group #2** Darker than control. Lighter than group 2	**Exp. Group #3** Darkest color product

Constants

Constants are the factors within an experiment that are kept the same for all groups or trials in an attempt to reduce the influence of additional variables. Once you have chosen the independent variable, you must design an experiment to take all of the other potential independent variables into account and make them constant. Otherwise, you will not be able to support a clear relationship between the two variables for which you have data.

How you decide to perform the experiment, meaning the step-by-step procedure, is crucial and can greatly influence the integrity of your experiment. Your treatment of each of the groups must be the same in every way. When analyzing the data after the experiment, you will have to critique your methods to see if something you may have done, or failed to do, influenced the results. This is another reason why background research before starting the experiment is so important.

You need to consider what it means to provide a constant environment for all the groups. For example, in a plant experiment where different intensities of light are used, it is likely that the soils will dry out at different rates. Does keeping the water a constant mean that each plant gets watered the exact amount, on the same days of the week? Or does keeping the water a constant mean that each plant gets enough water so that its soil is moist 1 cm below the surface? Although there is not always a "right" answer to these types of questions, you need to do background research to determine which methods would introduce the least amount of error.

In the reproduction/temperature experiment on planaria (pp. 17–18), the constants might include the methods used to observe and handle the planaria, the length of time each group receives light, and how often planaria are fed and environments cleaned. The list of conditions to keep constant within your experiment can be extensive. It is important to learn as much as possible about the entity being studied AND about the independent variable. You want to be as informed as possible about any additional factors that may influence the results.

Be careful that by controlling for one extraneous variable you are not introducing another one. An *extraneous variable* is a variable in addition to the independent variable that may influence the results of an experiment. Extraneous variables can introduce errors if they are not controlled or significantly decreased. For example, if planaria specimens are placed in different rooms to keep the varying temperatures from interfering with the experimental and control groups, additional variables have now been introduced. The different rooms might have varying amounts of light or might be used more or less frequently by people. You will not be able to control everything, but you will have to make decisions on what is least likely to influence the results. Be ready to address any limitations of the experiment. Explaining the efforts that went into reducing the effects of extraneous variables is important. Figure 2.4 shows a sample of what a biological experimental design might look like when put into an experimental design table.

Difference Between Quantitative Data and Qualitative Data

Quantitative data are data that use numbers with a unit of measurement—for example, the length of an insect in millimeters or the weight of a projectile in kilograms. *Qualitative data* are data that describe characteristics or qualities, such as color, odor, or texture, or data that describe category frequency or ratings, such as stem sturdiness (e.g., "sturdy," "somewhat sturdy," "limp"). Therefore, both describe the same situation but in different ways. While quantitative measurements are of upmost importance in all STEM-based research, qualitative descriptions of data are appropriate to supplement and give a different view of the same data.

> *Quantitative = data that can be expressed in numbers (quantified)*
>
> *Qualitative = descriptive data or data that has been put into categories (i.e., categorized data)*

Quantitative Data Uses

Quantitative data are the primary data collected for most STEM research. The purpose in collecting quantitative data is to enable you to categorize, organize, and classify your observations in such a way that the experimental groups can be compared mathematically to one another and to the control group. In other words, the quantitative data collected throughout the experiment (hourly/daily/weekly) can later be calculated into changes over the course of the experiment to determine if the difference is statistically significant. End of the experiment mathematical calculations may include means, modes, medians, total change, rate of change, or speed of change. These numbers are used to determine whether the differences are statistically significant. See Chapters 7–9 for a lot more information about mathematical analyses.

As the researcher, you will first consider quantitative measurements by measuring the effects of the dependent variable. You must consider using

Figure 2.4

Sample Biology Research Design Table

Hypothesis Draft				
If the amount of solid surface on top of the soil is related to the strength of the seedling, then seedlings will break through thinner surfaces more consistently and with less damage to the seedling.				

Independent Variable	**Background Questions**			
Varying depth of solid surfaces for seedlings to grow through	What species of seeds would best be used? What type of seed has a fast germination rate and is easy to grow in controlled conditions? What are the best solid surfaces to use? (Plaster of paris, concrete mix, spackling paste?) What other variables might be introduced by using these materials? How can I reduce those? What are the best ways to measure "strength" of seedlings? (Crack of surfaces, speed at which they get through the surface?)			

Dependent Variable	**Constants**			
Quantitative \# of days it takes to break through surface width/length of the crack Thickness of seedling stem *Qualitative* Condition of the seedling during and after breaking through surfaces Conditions of roots and seedling	Seedlings all have the same lighting, watering, and feeding schedule (plants are rotated weekly). Data collection is done at the same time every day. Temperature of the room remains the same for all seedlings. Seeds of the same kind came from the same package. Seeds are all planted in the same type and size container (clear plastic cup). All seeds have the same quality and amount of soil underneath the solid surface.			

Experimental Groups and Control Group	**Control Group**	**Exp. Group #1**	**Exp. Group #2**	**Exp. Group #3**
	No solid surface (just soil)	.5 cm depth solid surface	1 cm depth solid surface	1.5 cm depth solid surface

mathematical measurements such as area, angle, conductivity, density, electrical current, force, heat, humidity, length, light intensity, mass, pH, pressure, salinity, temperature, time, velocity, volume, or others. Your teacher will most likely require you to use only metric units or the International System of Units (SI), so you will use *centimeters* instead of *inches* and *liters* instead of *gallons*. The quantitative data you choose to record should be based on your background research.

Be sure to have reasons that support the relationship between the independent and dependent variables. For example, in an experiment that seeks to determine whether powdered drinks contain different food dye concentrations, it would not make sense to measure pH because pH is not the factor being studied in the relationship between the variables.

Quantitative data are used primarily to measure your "effects." There are other ways, however, in which quantitative measurements are a part of your experiment. First, you need to record measureable differences between your experimental groups. For example, if your independent variable is pH, you must record accurate measurements to ensure that the pH levels for each experimental group are appropriately different from one another. Second, you need to record quantitative measurements to ensure that extraneous variables remain constant. For example, temperature is often an added influence in an experiment and so must remain constant. Therefore, you must find ways within the experimental design to keep temperature the same and then plan to monitor and record this measurement periodically throughout the experiment.

The type of data and how you collect it will depend on what type of STEM research project you plan to do. In mathematics, physics, and population and human genetics research, the data may already exist. You may even be able to find reliable resources of data online. Then, it is a matter of obtaining the data and organizing it so that an analysis correlating it to the dependent variable can be made.

Qualitative Data Uses

Qualitative descriptions help you record changes within your experiment that may not necessarily be measureable. When collecting these observations, you describe how something looks, smells, feels, sounds, or tastes (when appropriate) or categorize it into a specific category. However, just because you may not be using numbers, don't lose your objectivity. Your observations should be scientific in nature and not make judgments or inferences. An *inference* is a conclusion, based on facts, that is perceived to be true by the researcher. Be careful, however, when you make an inference. The statement

"The solution looks normal" is an inference, but this conclusion obviously is based on observations that are not recorded. Although you may know what you mean, inferences written without factual descriptions will not help you compare results at the end of the experiment. Statements that include inferences are best saved for after data are collected. Instead, the actual observations, which lead to the inferences, should be recorded. Remain scientific and use detailed and descriptive language.

If you have trouble determining how to describe qualitative data, ask yourself, "What is 'normal' about this?" Make a long list of adjectives to describe the qualitative aspects of your dependent variable. For example, if you are studying viscosity of a fluid, list words that will help you describe varying thicknesses of the solutions—for example, *stringy, thready, dense, clumps,* or *runny.* If you photograph the entities throughout the experiment, you'll be able to compare qualitative differences. You may notice something in photographs that you didn't notice on a day-to-day basis. These observations will help supplement the quantitative data that you collect.

In addition to narrative descriptions, qualitative data can also be in the form of category frequency or ratings, both of which use numbers. Counting frequencies allows you to keep track of changes that are not normally quantified. For example, to record color change, you could use paint swatches, with each gradient of color assigned consecutive numbers—perhaps low numbers for lighter shades and higher numbers for darker shades. In a catapult-testing experiment, after research and/or pretrials, you might determine that there are three basic arch shapes in which the projectile might fall. After each trial, you could measure distance (quantitative) but also determine which of the three arch categories a catapult belongs to (qualitative).

If you choose to do behavioral research, you might collect data on location, like at a zoo, for animal behavior or at a coffee shop for human behavior. Recording behavior is a good time to use qualitative data. Behavioral research can be recorded several ways; the most common are focal sampling, scan sampling, and sequence sampling (Morgan 2009).

- In *focal sampling,* you choose one individual or group of individuals and record your observations for a set length of time. You watch and record everything you observe, writing in a narrative form.

- In *scan sampling,* you record the activity of an individual or group at preselected time intervals. Scan sampling should give you a sample representation of the behaviors taking place, and if you predetermine categories, it will also allow you to tally behaviors that can be used in data analysis. For example, if using scan sampling in the river otter experiment (Figure 2.1), you might observe the river otter in two-minute segments for several hours. At the moment each two minutes has passed, you would record what the otter is doing. Otter behavioral categories to be tallied might include walking, swimming underwater, floating on back, diving, grooming, foraging, or playing. Scan sampling helps keep an accurate record of observed behaviors as well as a record of changes over time if multiple observations are made.

- In *sequence sampling*, you record behaviors that occur within a sequence, in the order in which they occur. The rubric in Table 2.1 is an example of sequence sampling. The rubric was designed for a horse-training experiment in which the researcher wanted to keep track of a horse's progress as it learned a new skill. The behavior (taking a first step) was broken down into smaller pieces and then used during each training session to record the progress of the horse as it learned the new behavior.

Table 2.1

Sample Rubric for Observation of Horse Behavior (From Lifting the Hoof to Taking First Step)

1	2	3	4	5
Slightly bends knee for less than a second and puts back on ground	Bends knee but leaves toe of hoof on ground for about a second and then puts weight back on ground	Lifts hoof off ground but puts it back down less than a second	Lifts hoof off ground for a second and then puts weight back on ground	Lifts hoof off ground for more than a second

Table 2.2 shows sample quantitative and descriptive data for various types of research projects. The table may help you tell the difference between quantitative data and descriptive data that you collect during your experiment.

Table 2.2

Sample Quantitative and Qualitative (Descriptive) Data for STEM Research Projects

STEM Field	Entity Studied	Quantitative Data	Qualitative (or Descriptive) Data
Anatomy and physiology	Elbow joint range of motion (ROM)	110°	Patient winced at 95° but was able to go to 110°.
Biology	Earthworm growth	62 segments above the clitellum, 170 below	Worm pink at the posterior end and brown everywhere else.
Chemistry	Precipitation reaction	2.1 g	Bright yellow precipitate = 4 (scale 1–5)
Geology	Soil porosity	360 ml	Color of the soil did not change with the addition of water, but a bitter odor was noticeable.
Mathematics	Travel time	40 min.	The roads were wet because it was drizzling at the time of data collection.
Physics/engineering	Tensile stress and strain	87 g 6.4 cm	Cord made loud cracking sounds before it snapped.

Chapter Questions

1. What are the main components of an experimental design?

2. What is the purpose of having a hypothesis in a STEM-based research project?

3. When doing background research on independent and dependent variables, what sort of information will help you write a good research design?

4. How do the individual entities or trials within the experimental groups differ? How are they the same?

5. How are the constants different from the control?

6. How should the consideration of extraneous variables affect the design of a STEM research project?

7. How do quantitative data differ from qualitative data?

8. Why might recording inferences (instead of facts) interfere with data collection?

Chapter Applications

Now it's your turn. Look back at your Student Handout #1, Focusing Preliminary Research Ideas, on page 14. Use it to complete the Student Handout #2, the Research Design Table, on the following page. Refer to the example design tables provided in this chapter. Carefully consider the elements of your experiment. Think about the variables that can best be observed and measured, taking into consideration the equipment, resources, and lab skills you have at your disposal. Consider advice and suggestions from your teacher and your classmates. Completing this table in writing will help you determine the strengths and weaknesses of your research design. It will tell you what you still need to learn more about. Don't be surprised if you complete several drafts of the table, maybe on completely different topics.

If you are working with a group or with a partner on this project, your teacher may prefer that you brainstorm together regarding the research ideas and variables and then complete the remaining parts of Student Handout #2 individually. In that case, group members can discuss the differences between the different proposed research designs and then combine the best of each version to make a single group draft. At that point, consider typing up the group draft and posting it to a Google Doc that you can share with all members of your group and with your teacher (for more information on Google Docs, see p. xxviii). In that way, each group member can make edits to the document, and the teacher can check on the group's progress.

The next chapter will help you develop research questions to help focus your background research. Continue doing background research on your topic. Though it may seem contradictory, the more background information you have, the better you will be able to modify your research design.

References

Cothron, J. H., R. N. Giese, and R.J. Rezba. 2006. *Science experiments and projects for students: Student version of students and research.* Dubuque, IA: Kendall/Hunt.

Filson, R. 2001. In search of ... real science. Retrieved March 4, 2011, from Access Excellence: *www.accessexcellence.org/LC/TL/filson.*

Gordon, J. C. 2007. *Planning research: A concise guide for the environmental and natural resource sciences.* New Haven, CT: Yale University Press.

Morgan, K. 2009. Notes on behavioral recording techniques. Retrieved March 15, 2011, from Wheaton College website: *www3.wheatonma.edu/kmorgan/Animal_Behavior_Class/ recordingmethods.html.*

Research Design Table

Name _____ Class _____ Date _____

Directions: Complete the following table with your research project idea.

Hypothesis Draft

Independent Variable	Background Questions

Dependent Variable	Constants
Quantitative	
Qualitative	

Experimental Groups and Control Group	Control Group	Exp. Group #1	Exp. Group #2	Exp. Group #3

—3—

Background Research and Note Taking

Introduction

Now that you have organized some basic ideas of your experiment into a research design table, you may have an idea of what your experiment might look like. Although you have already done preliminary research to get conceptual ideas for your experiment, the next step is to write more formalized background research questions with the goal of learning your topic inside out. This chapter will help you take the information from your research design table, which you completed in Student Handout #2, and categorize that information into background research questions. This chapter will also provide tips on how to identify reliable resources, introduce different methods of note taking and documentation, give suggestions on how to avoid plagiarism, and describe online tools that will help you during the background research phase of your project.

Learning Objectives

By the end of the chapter, you should be able to

1. summarize how background research questions are used to organize note taking,

2. describe background research methods that will give you reliable resources,

3. determine the aspects of what must be organized when conducting background research,

4. explain how to avoid plagiarism, and

5. describe technology tools that increase efficiency of the research process.

Key Terms

Documentation: The practice of referencing or citing previous works within a piece of writing according to an "official" documentation style such as that of the Modern Language Association (MLA) or the American Psychological Association (APA).

Scholarly research: Research and writing performed by an academic (e.g., a *professor*— a person who does research and teaches at a university or college) that is usually based on original research or experimentation.

Writing Background Research Questions

Background research questions must be written in such a way as to cover everything you need to know to conduct the experiment. These questions help you focus on what you really need to know. Research questions are to be written in a general way, not in a way that could be answered in one or two sentences. For example, the question, "What do river otters eat?" is not the best way to word a background research question because it can be answered with just one idea. Although that question is important and needs to be answered to do the experiment, background research questions are more general—for example, "What is needed to care for river otters in captivity?" This research question allows you to answer questions about what otters eat *as well as* what they need to sleep, drink, and swim. It also may bring up environmental issues that you may not have considered.

You should have at least one background research question for each of the following four categories:

1. Entity

    o Specific types that are easily studied

    o Its structure and function

    o Handling/care/safety/ethics within a controlled environment

2. Independent Variable

    o Its structure and function

    o How it can safely and ethically be manipulated

3. Dependent Variable

    o Its structure and function

3

o How changes can best be measured, recorded, and observed

4. Connections Between the Entity and the Independent and Dependent Variables

  o Learn what is already known about these relationships (i.e., previous research on the topic)

Your background research questions for the river otter study might be something like these questions:

1. Entity

  o What is known about this specific type of river otter?

  o How do zookeepers at your local zoo care for river otters? (Include the genus and species of river otters you saw at the zoo and address safety and ethical issues as part of this question.)

2. Independent Variable

  o How do weather conditions change air temperature and water temperature?

3. Dependent Variable

  o What are common behaviors of river otters? When and why do they behave this way?

  o What methods of observation can be used to record otter behavior? (How can I record otter behavior in my lab notebook?)

4. Connections Between the Entity and the Independent and Dependent Variables

  o How does river otter behavior change as the temperature decreases?

  o How has this topic been studied in the past?

A research project should have between four and six research questions. If you have more than six questions, your questions are likely to be too specific and need to be consolidated. You will probably have two questions about your entity, two about your dependent variable, and one each for the independent variable and the connections between the variables. Don't be discouraged if, in your research, you don't find much, if any, information specifically about connections between your entity and the variables. That is why you are doing the research project—to look for those connections.

Once the members of a group have written the first draft of background research questions, the draft can be uploaded to Google Docs so that all members have access to it. Your group may decide to distribute different background questions to each group member. That way, time is used efficiently with each member looking for answers to different background questions.

Starting Background Research Early

Although you might think that these introductory steps to the background research process will not take long, think again! Doing the preliminary research, developing good research questions, and writing a high-quality proposal are probably the most challenging parts of doing experimental research. Turning a good research idea into a doable research experiment is not a simple process. You must spend time doing background research to understand the problem you want to address. The time you spend now will make the experimental and writing processes much easier.

(You'll notice that we haven't had a discussion about hypotheses yet (except briefly on pp. 17–18). It is important to become very knowledgeable about your topic *before* writing a hypothesis. You need to know enough to make an educated prediction of the outcome of your experiment.)

Use Student Handout #3, Background Research Questions, on page 54, to begin developing these questions. It is important that your teacher approve these questions. Don't take your teacher's suggestions as criticism. Instead view the teacher approval process as a way to make sure the time you spend researching is well focused.

Identifying Reliable Scientific Resources

To answer the background questions you have developed for your research topic, you'll most likely be looking for resources at a library and online. Locating and obtaining resources from a library has traditionally been the way to determine whether or not a resource is reliable (i.e., accurate and trustworthy). Printed materials at libraries are usually reliable because they have been through an editing and publishing process. Many of these printed materials now have identical online versions that are available through libraries' paid subscriptions to specific databases (from which full-text versions can be downloaded).

For a research project at this level, you are expected to include scholarly research resources. *Scholarly research* is writing done by an academic (i.e., a professor or some other teacher or researcher at a college or university) that is usually based on original research or experimentation. Scholarly research, usually published in research journals, is highly respected because the writing had to be peer-reviewed by experts in the same field. This type of writing will be heavier in STEM-specific vocabulary than other resources you read on your topic, particularly nonscholarly writings on the web. Most of the scholarly research articles are protected behind firewalls and require a connection via a library. The library connection gives you access to high-quality, full-text scholarly articles, which you can e-mail to yourself. You'll also find citations for journal articles that the library may have in paper formats.

To gain access to library databases, you must either be on campus or have login permission. Your high school library and local public library may subscribe to certain databases that will give you access to some scholarly writings, so ask your librarian for help. University and college libraries may grant visitors occasional login privileges, even if you are not a current student. If going alone to a university library, call the library ahead of time to ask about its policies. Most university libraries do welcome visits from high school classes doing serious research through field trip arrangements made by teachers, school librarians, and/or school administrators. These visits benefit both you and the university library. You get to do college-level research early and to experience a nearby university's research services. The university gets to host you as a potential future customer and show you a little of what college life is like. Although as a guest at a university you may not be able to check out paper resources, you will be allowed to make photocopies and e-mail full-text articles to yourself.

There are also current and reliable materials that are available online that are not stored behind firewalls, but you will not necessarily be able to identify which ones are reliable if you use a basic search engine like Google or Yahoo. With a growing movement toward Open Access (OA), research institutions are increasingly making scholarly writing available for free. The sidebar provides an introductory list to databases that will help you identify free scholarly research.

Free, Open-Access Resources

- Directory of Open Access Journals *www.doaj.org/doaj?func=searchArticles*

- Google Scholar *http://scholar.google.com*

- InfoMine *http://infomine.ucr.edu*

- Infotopia *www.infotopia.info*

- Medscape Reference *http://emedicine.medscape.com*

- National Science Digital Library *http://nsdl.org/search*

- Public Library of Science *www.plos.org/journals/index.php*

- Scirus *www.scirus.com/srsapp/advanced/index.jsp*

- Open J-Gate *http://openjgate.org/Search/QuickSearch.aspx*

- Wiley Online Library *http://onlinelibrary.wiley.com*

- OAIster *http://oaister.worldcat.org*

- WorldCat *http://worldcat.org*

Methods of Note Taking

At this point, let's take a step back in the process. Even before you begin searching for resources to answer your research questions, you should plan how to organize your note taking. Do not just start reading and jot down sentences or phrases. After you finish your experiment, you will be writing a scientific paper or preparing a poster, and therefore, you must share with the reader where the information came from. For everything you read, record not only the name of the resource your information came from but also the page numbers on which it appeared. To save you time and focus your research energy, you should also organize your notes within the categories of your research questions.

Brief Introduction to Documentation

The word *documentation* refers to the practice of referencing or citing previous works within a piece of writing in accordance with an "official" documentation style guide, such as that of the Modern Language Association (MLA) or American Psychological Association (APA). There are many documentation styles used by researchers; however, in this handbook we will only learn MLA. I'll go into the details of how to document a paper using MLA style in Chapter 10. Right now, you need to understand the basics of documentation so that you collect the appropriate information as you take notes from your resources.[*]

The most important principle of documentation is that you give credit to the ideas, information, or expressions of others in two places within your writing: in parentheses within the narrative of your text and in a reference section at the end. Perhaps up until now, you have been allowed to write a paper and simply list the resources at the end. This method is not sufficient for the level of writing research you are doing now. It is *imperative* that as you take notes you keep track of where the information came from and what pages the information appeared on. Consequently, you must remain organized while taking notes. To prevent yourself from inadvertently plagiarizing, you should get into the practice of writing your notes in short phrases, not in complete sentences.

---

[*] I chose to use MLA as the documentation style for this handbook because it is what most high school English teachers use. Also, MLA style is the one you will most likely encounter your freshman year in college. Note, however, that professional STEM scientific papers are never written in MLA style. What is important is that you learn the principles behind documentation. In college, if you decide to enter a STEM field, it will not be difficult for you to transfer what you have learned from MLA to another style of documentation. If you plan to enter your research at a national symposium or fair contest while you are still in high school, be sure to refer to their guidelines regarding documentation style. Most competitions do not require a specific style—only that it is applied consistently and correctly.

Organizing the Note-Taking Process

There are many note-taking methods. I'll discuss two: using note cards and using a notebook.

There are also free online note-organizing tools that allow you to create accounts, format citations, take notes, and share the citations and notes with other individuals. These sites include but are not limited to the following:

- EasyBib (*www.easybib.com*) allows you to create virtual note cards, associate them with specific sources, and add page numbers and tags so you can easily find information you have recorded.

- NoteStar (*notestar.4teachers.org*) allows you to create subtopics to match your background questions, take notes, and organize notes and sources. It can be used in conjunction with (*thinktank.4teachers.org*).

- SpringNote (*www.springnote.com*) is a specialized wiki that helps you organize your notes online by setting up either a personal notebook or a group notebook.

- Noodletools (*www.noodletools.com*) includes links to online databases to help you find reliable resources. It also functions as a place to record information and organize online note cards. In addition, it will support you in formatting your resources into MLA style.

Even if you plan on using an online note-taking website to organize your research, you should still read the remainder of this chapter so that you understand how your notes should be organized to make the scientific paper or poster easier to write.

Note-Card Method of Organizing Background Research

1. *Research Question Cards:* Assign each background research question a number and write this number and question on a single note card (see Figure 3.1, p. 42). Consider using different-color cards for each question.

Figure 3.1

Sample Research Question Card

Place the assigned library research
question number in the upper left corner.

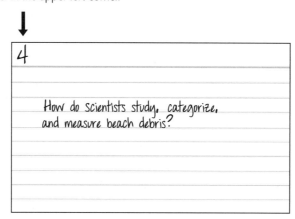

2. *Bibliography Cards:* For each resource you use (book, journal, or website), you will write the bibliography information (using MLA style—see Chapter 10) on a single card, assigning a letter to these resources as you go (see Figure 3.2).

Figure 3.2

Sample Bibliography Card

Place the assigned letter
for this resource in the
upper right-hand corner.

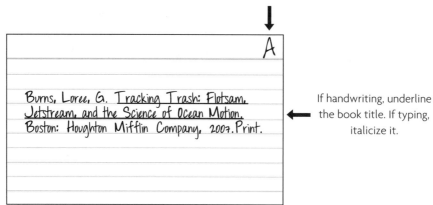

If handwriting, underline
the book title. If typing,
italicize it.

3. *Note Cards:* Begin taking notes on the note cards (see Figure 3.3).

   a. In the upper right-hand corner, put the letter of the resource from which the information is coming.

   b. In the upper left-hand corner, put the number of the background research question the notes are addressing. If you want to color code your note cards, coordinate the color with the research question card—not the bibliography cards!

   c. In the lower right-hand corner, put the page numbers where you found the information. When recording notes from pages of actual books, journals, and PDF files, use the abbreviation "pg." When recording notes from a web page, do not use page numbers; instead include paragraph numbers and use the abbreviation "para."

Figure 3.3

Sample Note Card

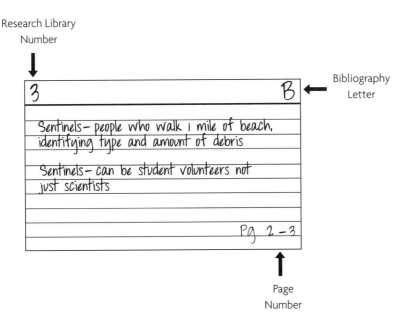

As you take notes, it is quite possible that one resource will help you answer more than one research question. Be sure to start a new note card when the question you're addressing changes. Each note card does not need to be full.

In favor of the note-card system: *Once the note taking is complete, the cards can be shuffled and reorganized in a way that makes writing the paper easier. (The keys to making this system work are to keep the cards organized and to carry them around without misplacing them.)*

Against the note-card system: *Loose note cards are easily misplaced or lost. Consider hole punching the cards and using metal rings to keep them together. Recipe boxes with tabs to separate the research questions also work well.*

Notebook Method of Organizing Background Research

This method uses the Notebook Organizer provided at the end of this chapter (pp. 55–56). This system also works well if you prefer to take notes electronically or in a spiral-bound notebook or composition notebook (where pages won't fall out).

1. *Bibliography Information Page:* Make the front pages of your notebook the place to list your resources (books, journals, or websites). You list these resources (using MLA style; see Chapter 10), assigning letters to them as you go (see Figure 3.4).

Figure 3.4

Sample Bibliography Information Page

	Bibliography Information
A	Sheavly, Seba B. *National Marine Debris Monitoring Program: Lessons Learned.* Virginia Beach: U.S. Environmental Protection Agency, 2010. Web. 4 Nov. 2010
B	Fradin, Judy B., and Dennis B. Fradin. *Hurricanes: Witness to Disaster.* Washington D.C.: National Geographic, 2007.
C	Burns, Loree, G. *Tracking Trash: Flotsam, Jetstream, and the Science of Ocean Motion.* Boston: Houghton Mifflin Company, 2007.
D	

> Notice that the entries are NOT in alphabetical order. That is because you will write the resources in the order that you find them. Later when you type the resources into your Works Cited, an alphabetical listing is necessary.

2. *Research Question Pages:* Write the number you assigned to each background research question on top of its own page in your notebook (see Figure 3.5).

a. If you are using the provided Notebook Organizer (pp. 55–56 have at least two to three pages for each research question.

b. If you are organizing this in a notebook, evenly divide the notebook pages between your research questions, so that there are plenty of pages to take notes for each research question.

c. If you are organizing notes electronically, insert page breaks between each research question, and consider using tables to organize the note-taking described in step 3, below.

Figure 3.5

Sample Research Question Pages

1. What sorts of trash are found in the ocean?

2. How does the direction a beach faces affect the kinds of biotic and abiotic items found on the beach?

3. How do the water currents at my local beach affect debris brought in from the ocean?

4. How do scientists study, categorize, and measure beach debris?

5. What does the available data say about the type of debris found on beaches facing in different directions?

3. *Taking Notes:* Begin taking notes in your notebook (see Figure 3.6, p. 46).

a. Place the bibliography letter in the first column of the table

b. Place the page numbers in the second column.

- When recording notes from pages of actual books, journals, and PDF files, use the abbreviation "pg."

- When recording notes from a web page, do not use page numbers; instead include paragraph numbers and use the abbreviation "para."

Figure 3.6

Sample Note-Taking Entries

Research Question (and Number):

4. How and why do scientists study, categorize, and measure beach debris?

Bib Letter	Page #	Notes: In Your Own Words!
C	4	—study debris: if large amounts of one item (toys, sneakers) are found...scientists try to find the source
C	5	Large spills: sometimes containers that fell off cargo ships
C	4	"Tracking toys and sneakers gives us a chance to see what the ocean does with our trash...and we can learn from it"
C	5	Beachcombers: people who collect things they pick up at the beach...scientists communicate with them to get data
C	40	Sentinels —people who walk 1 mile of beach, identifying type and amount of debris (on paperwork provided) — can be student volunteers not just scientists
A	4	Monitoring beach debris can show the condition of the water
A	4	National Marine Debris Monitoring Program (NMDMP): developed a standard way to record beach debris
A	5	Goal of NMDMP: determine 1) is the amount of debris changing? 2) what are the major sources of debris?

> Notice: Notes are in short phrases, not in full sentences.

> If you are handwriting your notes, feel free to use more than one line for the notes. This makes finding the information much easier.

In favor of the notebook system: *Taking notes on notebook paper is what most students are used to doing, and notebook paper is less likely to be lost than note cards. This method organizes all the notes by research question, so that they can be further organized into paragraphs.*

Against the notebook system: *Once the notes are written in the notebook, you cannot change the order in which they were taken (unless you choose to enter your notes electronically).*

While reading through your resources and taking notes, do not become focused on answering only one research question. A single source may help you answer several background research questions. Therefore, while you read from one resource, you may take notes on three different pages (or several note cards) because there is information that answers different background research questions.

Avoiding Plagiarism

Plagiarism is "using another person's ideas, information, or expressions without acknowledging that person's work" (MLA 2009, p. 52). Although you probably understand that copying word-for-word from a resource and not giving credit to the author is plagiarism, what you may not realize is that *just using someone's ideas without giving credit to them in parentheses within the paper is also plagiarism.* Whether plagiarism is intentional or not, it is a serious offense.

There are two ways you can prevent yourself from unintentionally plagiarizing.

Avoiding Plagiarism When Taking Notes

The best way to avoid plagiarism while taking notes is to take notes in your own words. One way to do this is, first, read a paragraph; second, close the resource (or minimize the program on your computer); and third, from your memory, write your notes. Write short phrases that summarize rather than complete sentences. The idea is to give an accurate presentation of the author's ideas. Then recheck your notes to be sure that you have correctly interpreted the content as well as the author's intent. Be careful not to change the meaning of what the author intended to say. *An important reminder:* Keep careful track of where information comes from and its page number; this will save you a lot of time in the long run.

When taking notes, don't say to yourself, *I'll just copy this into my notebook now but I'll rewrite it in my own words later.* You run the risk of forgetting to rephrase the direct quotes and then you may be accused of plagiarizing. In the end, it is much better to have a paper that—though perhaps not very eloquently written—is your own than it is to have a paper with parts that have been plagiarized. If you do come across a set of words that you cannot rephrase without losing the meaning or tone, be sure to put quotation marks around the words in your notes, so that you can properly credit the author when using those words in your paper.

Avoiding Plagiarism When Writing the Paper

If you took your notes in your own words, avoiding plagiarism while you write the paper is much easier then if you copied from your resources—or even worse, if you skipped the note-taking process altogether. When you are ready to write the paper, understand that even if you changed an author's wording into your own, you still must document *within your paper* where the idea came from; it is not enough to just add the resource to the reference list at the end of your paper. Changing the author's words into your own *is not* enough to keep you out of plagiarism trouble. And, again, if you use an author's wording exactly, you must use quotation marks.

Another way to be sure that you do not plagiarize while writing the paper is to use a minimum of two different sources to back up each new idea in a paragraph. Use the facts you have in your notes to write, in your own words, about the topic at hand, and compare your research to earlier research (which you cite in your paper). It is your job to collect what the experts say about various topics and then organize this information to introduce or explain your research study. In your paper, you will put together many facts from different sources in a way that has not been done before.

It should be mentioned here that electronically copying-and-pasting any author's work, including text, photos, images, or graphs, is plagiarism, unless that work is given proper documentation. *Anything* you put in your paper must be documented at the place in the text where you use it, in addition to being added to the reference list.

Using Quotations Within the Paper

Although you may view quoting an author's words exactly as the quickest way to get information into your paper, use this method sparingly. Save quotes for well-worded phrases that lose meaning when reworded.

The example below shows how a direct quote should be documented. Notice how the writer begins the sentence and allows the quote to finish it. Whenever you quote from a source, be sure to discuss the quote in your paper—do not assume it will speak for itself. Integrate the quote into the idea of your writing. Here is an example from a student research paper:

> Preston describes the amplification of the Ebola virus as the "virus convert[ing] the host into itself" (18) causing the person to become depersonalized. Essentially, "the who" of the host "has already died, while the "what" of the host continues to live (19).

Notice that the author's name is integrated into the text before the quote and that the page number is given in parentheses after the quote. If you do not include the name of the author as part of the text, put it inside the parenthesis in front of the page number. (See Chapter 10 for more detailed information about MLA style.) If the person who is reading your paper wants to learn more from this resource, the in-text citation gives enough information to identify it in the reference list, or what is called Works Cited in the MLA style of documentation. The author cited—Preston—can be found by looking at the Works Cited alphabetical listing under P. Therefore, the following entry would be found on the Works Cited page:

Preston, Richard. The Hot Zone: A Terrifying True Story. New York: Anchor
Books, 1994.

No matter what method you use to organize your notes, be sure to pattern your organization on the Notebook Organizer on pages 55–56. You need to keep track of both bibliography information and page numbers.

Technology Research Tools

Conducting Searches Within Databases

It is important that you learn how to perform effective Boolean searches (for a discussion of the meaning of *Boolean*, see *www.internettutorials.net/boolean.asp*) within the databases you are searching. Using the operators *and*, *or*, and *not* will help weed out search results that are not relevant to your topic. As you search within online databases, don't always limit your searches to "full text." Since you are working on a long-term project, if the library has neither an electronic (PDF), nor a paper version of the article, you can request it through an interlibrary loan (ILL) program. In a matter of days, the article may be in your hands.

RSS Feeds and Readers

Ask your teacher what sorts of online resources will be accepted as credible for your STEM research paper. If you use the open-access databases listed earlier in the chapter, your resources are more likely to be credible. While scholarly journal articles are important sources of information, do not overlook a blog posting from a world-renowned scientist from Yale or a podcast of a mathematician presenting at an international conference. Digital media allows us to get information as it happens. Having said this, the amount of information available on the internet can be overwhelming. However, there is help.

The answer is using RSS feeds and readers. RSS feeds, which stand for Real Simple Syndication, allow you to subscribe to relevant websites and specific engine searches to have the information sent to you—all in one place. These news feeds, or aggregators, allow you to have news and pertinent topical readings come to you via your RSS reader to collect websites publications on topics that interest you. The first step is to set up a mailbox, or an RSS aggregator, known as an RSS reader. There are many to choose from—Google Reader, FeedBurner, and NewsGator, to name a few. Once you create an account, you scour the internet for places that have, or might in the future have, relevant articles, interviews, or scientific studies. While searching on the web, look for the RSS logo or a link that invites you to subscribe to a feed. Many sites allow you to have only certain topics forwarded to your RSS reader, not every post from the website. For example, the RSS feed at the *Scientific American* website allows you to request feeds on only specific topics or narrowed areas of science.

In addition to subscribing to specific news websites, you can also have the powerful search engines behind Google News (*http://news.google.com*) or Yahoo News (*http://news.yahoo.com*) update you when new web pages on your topic are published. When you go to either of these sites, use the advanced search, and work at narrowing and perfecting a search on your topic. Once you have a search result where most of the topics look pertinent, right-click on the "RSS" link or icon, and select the phrase "Copy Shortcut" (or "Copy Link Location" or "Copy Link Address") and paste the address into your Add Subscription form in your RSS reader. Now any new news will be delivered to you! If after a week you find you're receiving too many results, delete the feed, and go back and narrow your advanced search to come from specific sources. Although media news resources are *not* considered reliable for scientific writing, they will help you get ideas, and their references can lead you to more reliable resources. You can find more details of how to fully use RSS feeds in Will Richardson's book, *Blogs, Wikis, Podcasts, and Other Powerful Web Tools for Classrooms*.

Social Bookmarking Tools

For a long time, what was missing from reading information on a computer screen was the ability to highlight text, handwrite notes in the margins, and "dog ear" the page (I don't condone that practice, by the way). Well, social bookmarking sites now allow you to do all of these things AND share your notes and markings with others. Social bookmarking sites such as Delicious (*www.delicious.com*) and Diigo (*www.diigo.com*) allow you to highlight text with any color highlighter you want, add a virtual sticky note to a web page, and determine textual tags that will help you find the site later. Of course by doing all of this you have bookmarked the page. Think about the time you have wasted looking for a specific idea you know you had bookmarked on an extremely long web page. Imagine now that your highlighted text draws your eye to the correct place on the page and that your virtual sticky note reminds you how you thought that text might help you on your project.

The social aspect of these bookmarking sites is what makes them so powerful. The ability to share with others your bookmarks and the marks you made on the page is extremely beneficial. If you are working in a group of students on the same topic, you can create a group within the social bookmarking site and share the bookmarks that pertain to the topic with them. The members of your group can in turn make comments on the page. Now you have true collaboration—not to mention increased efficiency.

Electronic Mail

If you find an article by a specific individual, go look up him or her online. Maybe you'll find the author is an active blogger, or maybe you'll find more recent, or yet-to-be-published, articles. Most important, look for online contact information. If you have questions or would like suggestions from an expert, consider contacting him or her directly. The first level of contact should be an

---

E-Mailing a STEM Professional

What to Include in Your E-Mail

- Formal opening that would be used in a business letter: "Dear Dr. Smith,"

- How you found that person (what article or website of his or hers you read).

- Who you are, and where you are from (a student working on a research project from ABC High School).

- If your teacher has given you permission, mention his or her name and maybe your teacher's e-mail address so that the scientist can verify what you say.

- A clearly stated request.

Examples of Requests:

- Could you answer three questions for me?

- Could I interview you in a Skype (a "Voice Over Internet Protocol") conference call?

- I'm having a problem with my research design [explanation here]. How do you suggest I address this?

- Would you be willing to be my mentor throughout my project?

(continued)

E-Mailing a STEM Professional *(continued)*

- Your contact information.

- Sincere thanks, acknowledging the person's expertise and making sure the person knows you value his or her time.

- Sign off, as you would in a formal business letter: "Sincerely" and your full name.

What Not to Include in Your E-Mail

- Questions about the individual that you can find the answers to online.

- Casual language like, "Hey! I found you on the internet."

- IM/text-messaging abbreviations. (If you cannot find the time to type out full words, why should this expert spend any of his or her valuable time answering your questions?)

- General questions such as "How do you design an experiment on this topic?" The more specific your question, the more likely you are to get an answer.

- An attachment of your entire paper, asking for any suggestions.

---

e-mail. However, remember, you are speaking to an authority figure on your topic. You should spend a lot of time constructing the e-mail, use respectful language, and clearly state what you are asking of the individual.

Chapter Questions

1. How are background research questions used to organize note taking?

2. How can you be sure that your research methods will provide you with reliable resources?

3. When taking notes, what are the key parts that need to be organized?

4. What can you do to avoid plagiarism?

5. Chose one technology tool that you may consider using. How might it help increase your efficiency?

Chapter Applications

I trust that this chapter helped you determine which background research questions you will use to focus your library and online searching. Complete Student Handout #3, Background Research Questions, and have it approved by your teacher. If at any point during the research process you determine that you need to modify your research questions, do so only with the permission of your teacher. Determine which libraries to visit to gain access to databases. Also consider bookmarking the Open-Access websites provided in this chapter (p. 39) to have them available when you begin researching. Decide (or ask your teacher) which note-taking method (note cards or notebook) you will use. Familiarize yourself with this method, and purchase anything you need (note cards or notebook and anything else mentioned earlier in the Organizing the Note-Taking Process section, pp. 41–46) to get started.

If working with others on your STEM research project, have a meeting and determine the tasks that need to be accomplished during the background research stage. As a group you may decide to divide the research questions among the group members with the intent to have each member answer only the assigned question. However, if you organize your notes in Google Docs, you could assign group member resources, allowing each to answer any research question as it applies. Assign tasks, write them down, and have each group member sign the contract before turning it in to your teacher.

In Chapter 4, you will learn how to rewrite your hypothesis. The background research you are completing now will help you to develop a hypothesis that is testable and will move you forward to developing a more detailed research design.

References

MLA handbook for writers of research papers. 7th ed. 2009. New York: Modern Language Association of America.

Richardson, W. 2009. *Blogs, wikis, podcasts, and other powerful web tools for classrooms.* Thousand Oaks, CA: Corwin Press.

Background Research Questions

Name _____ Class _____ Date _____

1. [Entity] _____

_____

2. [IV] _____

_____

3. [DV]_____

_____

4. [Connections]_____

_____

5. _____

_____

6. _____

_____

7. _____

_____

Approved by Teacher _____ **Date** _____

Notebook Organizer

Name _____ Class _____ Date _____

Part 1: Bibliography Page

Using correct MLA documentation, write out the bibliography entries below. Use the letters below when taking notes.

A. _____

_____

B. _____

_____

C. _____

_____

D. _____

_____

E. _____

_____

F. _____

_____

G. _____

_____

H. _____

_____

Part 2: Research Question Page for Taking Notes

Research Question (and Number):

_____

_____

Using correct MLA documentation, write out the bibliography entries below. Use the letters below when taking notes.

Bibliography Letter	Page #	Notes: In Your Own Words!

— 4 —

Writing Hypotheses

Introduction

Now that you have done some preliminary background research on (a) your entity, (b) the independent variable, (c) the dependent variable, and (d) previous research completed on similar topics, you should have a better idea of what sort of experiment will help you address the questions you have about your topic.

As first noted in Chapter 2, the two purposes of the hypothesis are to formulate what you want to test and to define the limit of your experiment. You construct a hypothesis (in writing) *after* you do your preliminary background research but *before* the experiment begins.

Learning Objectives

During the course of the chapter you will

- write hypotheses that are testable by experimentation and

- write drafts of a hypothesis for your own experimental project.

By the end of the chapter, you should be able to

1. identify at what point—during the process of designing an experiment—the researcher should write the hypothesis and

2. explain the importance of a prediction in a hypothesis.

Key Term

Hypothesis: A tentative (i.e., not final and definite) and testable statement that proposes an explanation for an observable phenomenon.

Although "an educated guess" is a definition commonly given in response to the question, "What is a hypothesis?" this definition is not only inadequate but also misleading. The word *guess* suggests that a hypothesis is not based on any real background information but is just someone's hunch. In fact, hypotheses are written only after extensive background research has been done—often after scientists have made numerous, careful observations about a specific phenomenon. A better definition of *hypothesis* would be the following:

A hypothesis is a tentative (i.e., not final and definite) and testable statement that proposes an explanation of an observable phenomenon. A hypothesis can predict a possible connection between two variables within a phenomenon or event or it can predict a difference between two groups.

A hypothesis is "tentative" because it is a temporary statement that a researcher makes to test an idea. The researcher expects the statement to be either supported or rejected by the experiment. And even after a hypothesis is supported by one experiment, it does not make the connections between the variables "certain." Researchers recognize that there are variables in an experiment that they were not aware of and could not account for. Therefore, a researcher never says that a hypothesis has been *proven*; instead, it is *supported*. This is how the scientific process works.

A hypothesis can propose a possible connection between two variables within a phenomenon or event or it can predict a difference between two groups. This book focuses on the first type, which proposes a connection between two variables. If you are interested in a study that *compares differences* between two groups, such as comparing the effects of watershed disturbance on the plant diversity in two different areas, then you will need to learn more about research design that uses hypothesis-testing statistics. There are various online statistic tutorials, such as Stat Trek (*http://stattrek.com/Lesson5/HypothesisTesting.aspx*), that will explain how to write hypotheses for research studies that compare groups.

Writing Drafts of the Hypothesis

You have already chosen (in Chapter 2) the independent and dependent variables that you want to include in your hypothesis. Now you need to make a prediction of what type of relationship—either positive or negative—may exist between these two variables. For example, if you were looking to make a connection between number of hours studied and exam scores, you might propose a positive relationship:

As the number of study hours increases, test scores will also increase.

or a negative relationship:

As the number of study hours increases, test scores will decrease.

When writing a draft of your hypothesis, you might start by using any of the following sentence formats as a guide:

- If __(IV)___ is related to ___(DV)_____, then (predict the effect) .

- If the _(IV)_ is (describe the changes), then the _(DV)_ will (predict the effect).

- _(DV)_ will (predict the effect) when_(IV)_ (describe the changes).

Once you have a draft of a hypothesis, you can move parts of the sentence around so that it flows well. What is important is that your hypothesis includes three elements:

1. Independent variable

2. Dependent variable

3. A prediction of what kind of effect the independent variable will have on the dependent variable. Predictions usually include phrases that propose differences, such as *increased/decreased, higher/lower, more/less,* or *faster/slower.*

Sample Hypotheses

For an experiment testing the effects of water temperature on planaria reproduction, a hypothesis could be written several ways. (In the sample hypotheses below, the independent variable is underlined once and the dependent variable is underlined twice.)

1. If the <u>speed of planaria reproduction</u> is related to <u>temperature</u>, then planaria in lower temperatures will reproduce more slowly than those in higher temperatures.

2. If the <u>temperature of a planaria's environment</u> is lowered, then the <u>speed of planaria reproduction</u> will decrease.

3. A decreased <u>temperature of a planaria's environment</u> will decrease the <u>speed of planaria reproduction.</u>

What is important is that these predictions are based on background research that confirmed that a relationship between the two variables is already known, allowing the researcher to write a hypothesis that includes a prediction. In an experiment predicting the effects of tennis court surfaces on a tennis ball's rebound height, a hypothesis could be written in several ways.

1. *If a* <u>*tennis court's texture*</u> *is related to a* <u>*tennis ball's rebound height*</u>*, then rough-textured surfaces will* *decrease* *a ball's rebound height.*

2. *The coarser the* <u>*texture of a tennis court*</u>*, the lower* <u>*rebound height*</u> *a tennis ball will have.*

3. <u>*Tennis court texture*</u> *will decrease the* <u>*rebound height*</u> *of a tennis ball when bounced on rough court surfaces compared to smooth court surfaces.*

Note that all three versions of the tennis court hypothesis include a description of a relationship between the texture of a tennis court and rebound height of a tennis ball, as well as a prediction of the effect.

It takes a lot of work to write a really good hypothesis, even if the hypothesis doesn't contain many words. In addition to introducing the two variables, a hypothesis also must state what kind of data will be collected. In the tennis ball examples above, "rebound height" is the data being collected.

The major difference between the original basic research question you started with in Chapter 2 and the hypothesis you are now writing is that the hypothesis statement must be testable. *Testable* means that the wording of the hypothesis makes it clear how a test will be performed to connect the two variables. This following statement is *untestable:* "Tennis court texture affects tennis ball rebound height." This hypothesis statement is not testable because it does not indicate how you plan on supporting that claim. Your hypothesis should predict a specific relationship between the independent variable and the dependent variable. The data you collect will either support or reject the predicted relationship. Anyone who reads the hypothesis should have an idea as to what the experiment will measure, although they will not know how the data will be collected.

Read the following hypothesis. What scientific facts did the researcher have to know before he or she could write this hypothesis? (*See the answer at the bottom of the page *after* you have answered this question.)

---

* The researcher had to know the following before he or she could write this hypothesis:
  • The rate of transpiration can be measured by its condensation levels.
  • The size of the wavelengths of light is determined by the color of the light.
  • Changing the color of light is a way to control the wavelength each plant receives.

4

If the <u>rate of transpiration</u> is related to <u>wavelengths of light</u>, then exposing a plant (Philodendron scandens) to shorter wavelengths of light will produce less condensation.

Writing a testable hypothesis takes time. In fact, a hypothesis may change quite a bit from when you first start working on it until you have a finished, testable hypothesis. Table 4.1 shows a specific example of how a hypothesis may progress over the course of its development.

Table 4.1

Progressive Drafts of a Hypothesis

Four Drafts of a Hypothesis	Comments on the Hypothesis
First draft: Plants and gravitropism (movement of plant stem and roots due to gravity)	Basic topic. Only the entity and a possible independent variable are given.
Second draft: The amount of root cap on a root affects plant gravitropism.	Describes the connection and is more specific than the hypothesis above but still not testable.
Third draft: Roots of plants that have root caps removed will differ in root gravitropism from plants that do not.	Proposes a connection between the IV and DV. This is not appropriate for an experiment because this relationship is commonly known.
Fourth and final draft: If root gravitropism is related to the amount of root cap removed, the more root cap cells are removed, the more the root will show change in angle growth.	Complete hypothesis. Proposes a connection between the IV and DV and tells what the experiment is testing and what will be measured. From this hypothesis, it is clear that the angle of bending roots will be measured.

As you take additional background research notes, get your proposal approved, begin to collect data, and then statistically analyze what happened, remember to frequently refer back to your all-important hypothesis. At the end of your experiment, you will be expected (by your teacher, mentor, or other people who are reviewing your experiment) to explain in your final research paper whether your hypothesis was supported or rejected and then explain why. More specifically, you will be expected to explain whether the relationship and prediction you made regarding the two variables was indeed supported by your data. Complete Student Handout #4, Practicing Writing Hypotheses, pages 63–65.

Chapter Questions

1. At what point during the process of designing an experiment should you write the hypothesis?

2. What is the importance of a prediction in a hypothesis?

Chapter Applications

Using Student Handout #4 compare the hypotheses written by other students in your class. How are they similar to each other? How are they different? Be ready to discuss which hypotheses are more easily testable and measureable.

Work with your teacher to write a testable hypothesis for your own research project. Have several versions of a hypothesis ready to share with your teacher and listen to his or her feedback. Because you have already begun background research, you should have a good sense of a prediction you can make in your hypothesis. Keep in mind that although hypotheses are not long in word count, they take careful thinking to write well. Share your several versions with your classmates and listen to their feedback. Continue working with your teacher until your hypothesis is accepted. In the next chapter, you will learn guidelines for writing your research proposal and enter into the current debate about the "voice" and grammar appropriate for a scientific paper.

References

Cothron, J. H., R. N. Giese, and R. J. Rezba. 2006. *Science experiments and projects for students: Student version of students and research.* Dubuque, IA: Kendall/Hunt.

Creswell, J. W. 1994. *Research design: Qualitative and quantitative approaches.* Thousand Oaks, CA: Sage Publications.

Farrugia, P., B. A. Petrisor, F. Forrokhyar, and M. Bhandari. 2010. Research questions, hypotheses and objectives. *Canadian Journal of Surgery* 53 (4): 278–281.

Gordon, J. C. 2007. *Planning research: A concise guide for the environmental and natural resource sciences.* New Haven, CT: Yale University Press.

Hollins, C. J. and V. Fleming. 2010. A 15-step model for writing a research proposal. *British Journal of Midwifery* 18 (12): 791–798.

Practicing Writing Hypotheses

Name _____ Class _____ Date _____

A *hypothesis* is a tentative and testable statement that proposes an explanation for an observable phenomenon.

Directions: For each of the following statements write two possible hypotheses. Underline the independent variable once and the dependent variable twice.

1. Electromagnetic fields have an effect on algae cells.

_____

_____

_____

_____

2. The shape of solar reflector cells has an effect on the amount of light collected.

_____

_____

_____

3. The local news station is more accurate in predicting the weather than the National Weather Service is.

_____

_____

_____

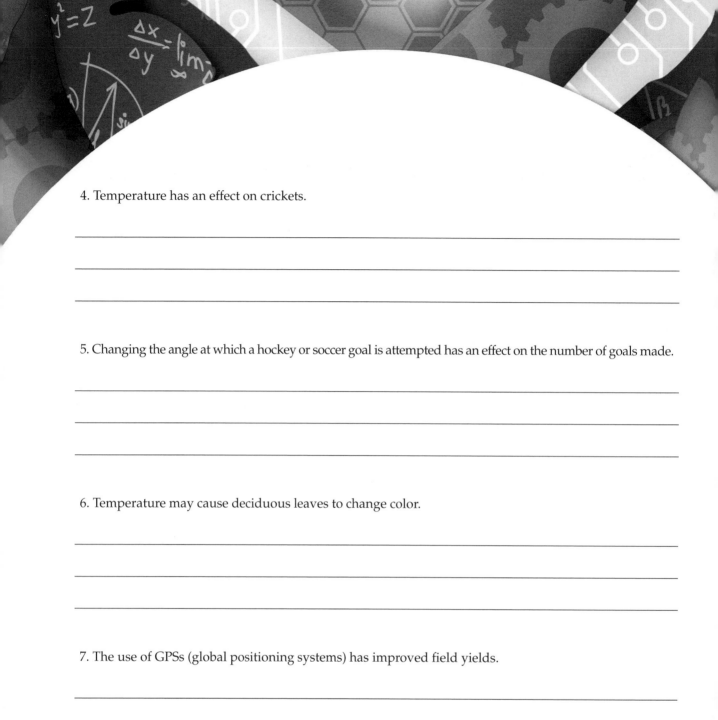

4. Temperature has an effect on crickets.

_____

_____

_____

5. Changing the angle at which a hockey or soccer goal is attempted has an effect on the number of goals made.

_____

_____

_____

6. Temperature may cause deciduous leaves to change color.

_____

_____

_____

7. The use of GPSs (global positioning systems) has improved field yields.

_____

_____

_____

8. The performance of various gasoline octane levels (87/90/91) differs.

_____

_____

_____

9. Is there is a statistical advantage to the team who scores first in a game?

_____

_____

_____

Now try writing a few hypotheses for your own research topic. Once you start to work with a hypothesis, you will discover the areas in which you need to do additional research.

_____

_____

_____

_____

_____

_____

_____

_____

_____

—5—

Proposal Writing

Introduction

After you have begun background research on your topic and have a preliminary hypothesis, you will write your proposal. A *proposal* is a document that describes the proposed experimental design of your research project. Essentially, it is a paper written by you to show your teacher that (a) you understand the background of your topic, (b) you have a testable hypothesis, and (c) you have an appropriate procedure to test that hypothesis. In the proposal you also address safety and ethical issues. The proposal will include the following six components: title, introduction, experimental design table, hypothesis, materials, and methods. If necessary, your proposal will also contain one or more appendixes.

Learning Objectives

The main objective of this chapter is to have you put together a proposal for your research project.

By the end of the chapter, you should also be able to

1. explain the purpose of writing a research proposal,

2. defend the need for a detailed methods section in a proposal,

3. describe how pretrials might apply to your own research project, and

4. analyze the disagreements that scientists and journal editors have about various aspects of scientific writing.

Key Terms

Pretrial: A pretesting of experimental methods with the purpose of tweaking the procedure before actual data collection begins.

Proposal: A written document that describes the proposed experimental design of a research project.

Depending on your situation, either your classroom teacher or a mentor will be the one to guide you through the proposal process. If you are working with an industry or university mentor, your experience is likely to differ from that of your classmates. For example, if you are working in your mentor's lab, the procedure of your experiment is probably already determined. However, it will be your job to fully understand and be able to explain the procedure adequately. That means spending time in the lab—acquainting yourself with the entity being studied, the equipment being used, and the background knowledge required to perform the experiment. Your mentor may be the one to review and provide feedback on your proposal or he or she may ask your teacher to do that. From this point on in this book, the term *teacher* will be used to represent either teacher or mentor.

Writing the proposal is a process. Do not view it as a one-time effort. Your teacher will provide feedback, most likely in the form of questions, that will help you to clarify—in your mind as well as on paper—what you intend to do. Then you will probably rewrite the proposal to address the issues that your teacher has highlighted. This exchange between you and your teacher may happen several times before your teacher gives his or her approval to the proposal and says that you can begin your research study.

The Proposal Components

Write the proposal paper with the following six headings: Title, Introduction, Experimental Design Table, Hypothesis, Materials, and Methods. (Your proposal might also contain one or more appendixes.) Follow these formatting rules for the headings in your proposal

- Center each heading on the page.

- Use Times New Roman or Arial font and 12 pt. type.

- Do not bold or underline the headings and or put any of them in all-capital letters—so, no **Hypothesis** or <u>Hypothesis</u> or HYPOTHESIS. Simply use: Hypothesis.

Title

List the working title for your research project. Titles should be concise, descriptive, and informative. They should be written in scientific style and therefore not in the form of a question. At a minimum, the independent variable and dependent variable must be included, but the more descriptive titles are, the better. Provide genus and species names of organisms and specific chemical names rather than vague references. So, for example, a better title than "Do Radish Seeds Prefer Acid or Bases?" would be "Effect of pH on *Raphanus sativus* Seed Germination."

Introduction

The purpose of the introduction section is to state the problem or topic and why you are studying it. You will explain how you are addressing the problem and how you plan to find a solution (Day and Gastel 2006). The introduction is more than just a "literature review" (i.e., a summary of all the journal articles you have read) or a collection of all the background information you have gathered. More information about how to write an introduction as well as how to properly document this section will be found in Chapter 10. (It is possible that your teacher may not want you to write an introduction as part of the proposal. If you do not write it now, you will write it later as part of the final paper.)

Experimental Design Table

Update your Research Design Table (p. 33) to match this proposal. That will allow your teacher to compare your intended research design with the methods you are now proposing. Your teacher will make sure that your procedure will actually test what your hypothesis is predicting.

Hypothesis

State the hypothesis that was approved by your teacher at the end of Chapter 4.

Materials

Provide a list of the materials required to conduct your experiment. Include supplies needed to set up the experiment, consumables (items that will be used up during the experiment), tools and instruments used for measurement, and any additional items you will use.

Include exact technical specifications for each material listed, such as purity of chemicals, their concentrations, and their suppliers; the type, brand,

and model of each apparatus; and genus and species along with characteristics such as age or sex of organisms. Be sure to give amounts (in metric units) when applicable. For example, instead of listing "water," list "600 ml of distilled water." The genus and species of any organism should be *italicized*. Furthermore, if an item is listed in the materials section, it must be listed in the methods section, and vice versa. It is easiest to write the methods first, then go back and list the materials.

Methods

The purpose of the methods section is to describe how the experimental design procedure will be carried out. This section should be written in enough detail so that another person would be able to replicate the experiment. The methods section also allows your readers to "judge the appropriateness of the experimental methods" and therefore the extent that the results can be generalized (Day and Gastel 2006, p. 60). The methods section describes the "how" and "how much" of the experiment and makes up the bulk of your proposal.

Before Writing the Methods Section

Before you begin writing the methods section of your proposal, you may need to do two things.

1. *Refer to your background research.* Read up again on various procedures that will help you obtain the measurements you are seeking. You may need to do more background research to find additional or supplemental procedures.

2. *Run some pretrials.* A *pretrial* is a pretest of experimental methods that you conduct to tweak the procedure before actual data collection begins. Pretrials will help you during the proposal-writing phase to describe in specific detail what you propose to do for your experiment. Pretrials also increase your chance of success with your experimental design because you can work out any kinks before collecting data.

   Running pretrials can mean different things depending on your research topic. In the case of a seed germination project, for example, the pretrial may involve trying several seed types before choosing a specific species. For a range-of-motion project, pretrials may require that you receive training on how to use a goniometer and practice taking measurements accurately so you can write a precise description of how the measurements will be taken. For a stream ecology project, you may practice various methods of recording water velocity to determine the best method for your experiment.

And for a behavioral study, you might develop and pretest categories or surveys to fine-tune them.

Using the Narrative Approach in the Methods Section

Traditionally there are two ways to write the methods section. Some teachers prefer that you write it as a step-by-step, numbered procedure—a set of instructions. Other teachers may prefer it in a narrative form (or an *essay*, as you would call it in English or history class). The narrative form is the one used in scientific journals. Both step-by-step and narrative method sections must include enough detail so that another individual would have no questions if he or she wanted to replicate your experiment exactly. If you are asked to write a narrative proposal, here are some suggestions for paragraph organization.

- Start the methods section with a paragraph discussing how the experiment will be set up. What preparation will need to be done before the experiment can be started? Describe any pretrials that were used to help determine the methods. Be sure to mention the number of entities being studied, how they will be labeled or organized, and the items and tools that will be used to structure the setup.

- The next paragraphs of the methods section should detail what is done on "Day 1" of the experiment. Day 1 is the first day that data are collected. In this description, use the scientific vocabulary you learned about your entity while doing background research. For example, when describing how the angle of a plant stem will be measured, use the technical scientific terms for plant structures, such as *internodes, petiole, node,* and *blade.* If the data collection methods will be repeated, you can tell the reader how often and with what experimental groups the methods will be used—for example, "The procedure will be repeated for a total of four trials for each amount: 0 milliliters, 15 milliliters, 30 milliliters, and 45 millileters."

Frequency of Data Collection and Number of Trials

When writing the methods section, you might ask (as many students do), "How often should I collect data?" or "How many trials do I have to do with each group?" The answers to those questions depend on two factors

1. *The variation in the entity being studied.* Organisms or objects that have a greater chance of variation in the data collected than other entities might require more trials or measurements taken for longer periods of time. For example, to account for sensitivity of protists,

you should use more organisms to begin with and as much data should be recorded as is reasonable. However, for a study comparing altered catapults, where there is less chance of variation in the data, you may need to collect fewer data points. However, the more data or trials you have for each level of the independent variable, the more confidently you will be be able to state the results.

2. *The speed at which data are available.* For your experiment, decide how quickly you expect changes will occur and plan your data-collection days accordingly. For example, for bacteria growth, you may choose to collect data every day (maybe even every 12 hours) for the first week. But for a slower-moving experiment, or one that includes trials that you determine, you may only need to collect data every two to three days to get appropriate data for your analysis. Your background research and pretrials should help you make this determination.

Describing Your Data Collection

Answer the following questions. Then, in your methods section, use your answers to describe your data collection.

- How will data be collected?

- How will measurements be taken?

- How will the data be measured or obtained?

- What tools and techniques will be used?

- How will qualitative observations be recorded (if applicable)?

- How often will data be collected and recorded?

- How long will the experiment last?

- How does your research design address potential extraneous variables (address each variable individually)?

After you have carefully described how data will be collected, describe the hourly, daily, weekly, or monthly tasks that you will need to complete to maintain the environment in which the experiment is taking place. What measurements or observations, such as temperature or humidity, will you record throughout the experiment?

You may be asked to rewrite the methods section several times until your teacher is confident that the research design has been developed enough to give you the best chance at testing what your hypothesis is predicting. The methods section of your proposal can be challenging, but it is a critical part of the experimental research process. Receiving repeated feedback from your teacher will keep you from having serious problems later when you are collecting and analyzing the results.

Appendixes

Appendixes (note that the preferred spelling for the plural of *appendix* is *appendixes,* not *appendicies*), which are found at the end of the proposal or final paper, can be any items that are too big to fit into the context of the paper, such as large tables, figures, survey questionnaires, or safety or ethical documentation.

If your research included using surveys or questionnaires, you must include as an appendix the actual survey questions that you plan to administer. If your experiment required that you submit documentation to an Internal Review Board (IRB), Scientific Review Committee (SRC), and/or Institutional Animal Care and Use Committee (IACUC), the signed approval documents or consent or assent forms themselves must be included in the appendix (see *www.societyforscience.org/isef/document).*

You may *not* begin any of your research until all forms have been approved by the committees. (See Chapter 1 for more detail about safety and ethical documentation.) In the methods section, be sure to refer the reader to the documents located in the appendix.

Scientific Writing

Students, young researchers, and even veteran researchers often believe that scientific writing should be filled with technical, terminology-filled jargon, long sentences, and the passive voice. None of these assumptions are true. Good scientific writing is concise and accurate and uses as few words as necessary. You should never write a sentence of 15 words if the same thought could be written in 9. For example,

> **Wordy:** *In the present report, the results of an experiment are described in which coffee and tea drinkers were tested to see whether ...*

> **Better:** *We tested coffee and tea drinkers to find out whether ...*

If technical references to equipment or scientific vocabulary cannot be avoided, do not assume that the reader knows everything you have learned; therefore, explain terminology the first time it appears in your paper. For example, if you used a centrifuge to do your experiment, explain what a centrifuge is the first time you refer to it.

Writing a scientific paper is actually more similar to writing a paper for English class than you might have thought. Many of the skills you have learned in English apply to your research writing, such as paragraph organization, construction of strong topic sentences, and proper spelling and grammar. However, there are some misconceptions of scientific writing that must be clarified.

The "voice" of research writing is a topic of dispute among scientists (Day and Gastel 2006). A sentence can be written in active voice or passive voice. In active voice, a subject performs the action. This type of sentence uses an active verb with a clear link to the subject. For example, "We [or "the researchers"] heated the solution to 44°C." The individuals (*we*) who performed the action (*heated*) are clear. In passive voice, the emphasis is put on what was done rather than who did it. The previous example written in passive voice could be, "The solution was heated to 44°C." *Who* performed the action is left out of the sentence entirely. In years past, passive voice was always used in scientific writing. According to scientists of that time, passive voice helped the reader focus on what was done, not on the individual who did it. (For that reason, the passive voice is sometimes still preferred for methods sections.)

Since the late 1990s, however, editors of scientific journals have come to prefer the use of active voice. This switch to active voice naturally led to the dispute of whether or not pronouns should be used in scientific writing. In the case of passive voice, the pronouns *I, us,* or *we* can be avoided. Active voice, on the other hand, requires that pronouns be used. Although some researchers prefer passive voice, others think it results in awkward sentences and a pompous tone. (For more on the different uses of active and passive voices go to *www.biomedicaleditor.com/active-voice.html.*)

Table 5.1 highlights the different types of writing and when each style is appropriate. Notice the use of second person is never acceptable in scientific writing.

Your science teacher will determine the voice you should use (active or passive) and whether or not you will be allowed to use pronouns. Researchers trying to publish their papers professionally must refer to an individual journal's author requirements before beginning to write because STEM journals have varying requirements for scientific writing. Many STEM-based journals indirectly advise authors to write in the active voice by referring writers to a style manual that supports active voice. Some journals' style guides ask that passive voice be used in the methods section but that active voice be used in all other parts of the paper. The style guides of the following professional associations (all having to do with science) recommend the use of pronouns and active voice as much as possible, but accept passive voice with the absence of pronouns as appropriate for methods sections: American Institute of Physics (AIP), American Chemical Society (ACS), American Medical Association (AMA), and American Psychological Association (APA).

Table 5.1

Use of Active Versus Passive Voice and of Pronouns in STEM Research Papers

Sample Sentence	Voice and Pronoun	Comment
"I will remove the ball bearing."	Active voice First person pronoun (Future tense)	Appropriate when writing a proposal.
"I removed the ball bearing."	Active voice First person pronoun (Past tense)	Appropriate when writing the final paper.
"The ball bearing will be removed."	Passive voice No pronoun (Future tense)	Appropriate when writing a proposal, especially in the methods section.
"The ball bearing was removed."	Passive voice No pronoun (Past tense)	Appropriate when writing the final paper, especially in the methods section.
• "You should remove the ball bearing." • "Remove the ball bearing."	Second person Directive	*Never appropriate.* These sentences are written as directions for someone else to follow rather than as an indication of what the researcher intends to do or has done.

Chapter Questions

1. What is the purpose of writing a research proposal? Who benefits from your writing of a research proposal?

2. Why must the methods section in a proposal be so detailed?

3. How might pretrials apply to your own research project?

4. Why do you think scientists and journal editors disagree about various aspects of scientific writing?

Chapter Applications

It's time to write your first draft of your proposal. You can expect your teacher to have you write the proposal several times until you have shown him or her that you have done adequate research, that your research design is appropriate for testing your hypothesis, and that you have taken appropriate safety and ethical issues into account. As you receive feedback, think carefully about the questions and comments posed by your teacher. Remember all suggestions are meant to help you sharpen your ideas as well as your writing.

If working with others on your STEM research project, hold a meeting to discuss group members' strengths and assign tasks—as you did before conducting background research. Document your discussion in writing and have each member sign. Then turn the document in to your teacher for accountability.

Once your teacher approves your proposal, the next step in the research process is to prepare a laboratory notebook. Chapter 6 shows you how to do that.

References

Day, R. A., and B. Gastel. 2006. *How to write and publish a scientific paper.* Westport, CT: Greenwood Press.

Every, B. Clear science writing: Active voice or passive voice? Retrieved March 17, 2011, from BioMedical Editor website: *www.biomedicaleditor.com/active-voice.html.*

International rules for precollege science research: guidelines for science and engineering fairs. 2010. Retrieved March 16, 2011, from Society for Science and the Public, Intel ISEF document library website: *www.societyforscience.org/isef/document*

— 6 —

Organizing a
Laboratory Notebook

Introduction

Congratulations! Now that you have written your research proposal and have a clear plan for your research project, you are ready to organize a notebook for recording the data you will collect from your research experiment. A *laboratory notebook* is an organizational tool that you will use to store and record all aspects of your experimental research project. I use the term *laboratory* in a general sense to include experiments that are done outside the laboratory setting. After all, the world is our laboratory!

Learning Objectives

The main objectives of this chapter are for you to create a laboratory notebook for your research project and to identify and construct data tables specifically designed for your project.

Also, by the end of the chapter you should be able to

1. explain the importance of keeping an accurate laboratory notebook and

2. explain the pros and cons of paper laboratory notebooks and the pros and cons of online laboratory notebooks.

Key Terms

Laboratory notebook: An important organizational tool for the researcher; used to store and record all aspects of the experimental research project, including procedures, data, statistical outcomes, graphs, and conclusions.

Raw data: The numerical data collected during an experiment before calculations or statistics have been applied to these numbers.

Purpose of Laboratory Notebooks

The laboratory notebook has three main purposes, which dictate its organization and its contents.

1. The laboratory notebook is where you store experimental research procedures, data, statistical outcomes, graphs, and conclusions.

2. The notebook is a place for you to record thoughts you have about the experimental design and any inferences you have regarding possible outcomes of the data you are collecting. In this sense, the notebook is a journal or a place where you talk to yourself about the experiment.

3. The notebook serves as an official record of the experimental project as a whole. Someone else should be able to pick up your notebook and (a) undertand how you performed the experiment and (b) be able to replicate it exactly.

Although there are software programs that help researchers organize their research projects (see pp. 79–80), most researchers still prefer paper versions of the laboratory notebook. In this chapter, I'll primarily discuss paper notebooks but will address technology adaptations as well.

Paper Laboratory Notebooks

Because the notebook is an offical dated record of the experiment, researchers put a strong emphasis on (a) using laboratory notebooks constructed in such a way that they reduce a researchers's temptation to tear out pages, (b) using permanent pens (i.e., those that do not bleed, such as pens you might use on ceramics), and (c) not using correction fluids or tapes. Buy a notebook where the pages are sewn into the binding. Composition notebooks work well and are inexpensive. "Official" laboratory notebooks are also available for purchase from companies like Book Factory, Scientific Notebook Company, and Amazon.

Do *not* purchase sprial notebooks or three-ring binders. It is too easy to remove pages that you think show undesirable or unfavorable data in light of your hypothesis. Removing evidence of procedural mistakes or data that do not support your hypothesis is highly frowned upon in the scientific community. The notebook should record everything regarding the experiment because you never know what may be significant. Therefore, never rip out any page of your laboratory notebook.

Because laboratory notebooks will include graphs and other items that may be printed from the computer, having a notebook large enough to accommodate these computer printouts is a necessity. The printouts will be glued or adhered to the pages within, so you will either have to reduce the font to make it fit or buy a notebook slightly larger than 8 ½ × 11 in. to accommodate these inserts (art supply stores are good places to find larger notebooks). In addition, be sure to find a pen that does not bleed when it gets wet. If you make a mistake when entering something into the notebook, just draw a single line through the word(s) and write the correct word(s). Don't use correction fluid or "obliterate the entry with an ink blob" (Purrington 2009, p. 33).

Online Laboratory Notebooks

Some scientists frown on the use of technology for a laboratory notebook. Two of the arguments against organizing and storing this important information electronically are (1) it is too easy to modify data or to remove undesirable results, and (2) the confidentiality and safety of the information are uncertain. However, because both wikis and Google Docs have a history function, any changes made to the web pages or documents are recorded, date stamped, and attributed to the person who made the changes. Therefore, data changed, added, or deleted is recorded. The confidentiality issue is a bit more complicated. However, advanced features of wikis allow them to be private and readable only by their members. Google Docs have various privacy settings as well, but one should be cautious and assume that anything posted to the web can be viewed by anyone. Overall, as long as individuals' names or personal identifying information is not recorded online, posting data collected as part of a high school project should not be a problem.

If you have consistent access to a computer with reliable internet service, you may want to consider having your laboratory notebook online. (Online laboratory notebooks also make sense to those of you working in groups. See the Data Collection Issues for Groups on p. 80 below for more information about this.) The purpose of the notebook is exactly the same as the paper version, but instead of attaching printouts to pages or directly recording data into a paper notebook, you organize, post, and link the same information online.

You first need to consider what online space will be the portal to your entire notebook. This most likely will be a web page. It could be a page you design yourself from scratch or it could be a wiki-like program. When constructing these pages, consider privacy issues and how much control you have over who can view your pages or, in the case of a wiki, who can make changes to them. (If you are designing a wiki and working in a group, all group members must have wiki accounts and be given special permission to make edits. It is likely that your instructor may also want editing privileges so that he or she can leave you comments and feedback.)

When organizing a laboratory notebook online, follow the suggestions given in the Components of a Laboratory Notebook section of this chapter (pp. 82–91). For starters, each section heading will be its own web page, with links easily accessible in a navigation area of the website. The keys to making a functional online laboratory notebook are (a) linking to appropriate pages within the wiki and (b) organizing links to additional notebook components logically.

(*Note:* The safety of the information posted online could be a concern. Just as you could lose your paper laboratory notebook, something could happen to the online laboratory notebook. You can reduce the likelihood of this tragedy by backing up the wikis and Google Docs files weekly.)

Data Collection Issues for Groups

If you are conducting research as part of a group, you will find that the data collection component has three challenges: (1) where the experiment will be conducted, (2) how to assign tasks to group members, and (3) how to organize a group laboratory notebook.

Determining the Location of the Experiment

First, you will need to consider where the experimental setup will be. It is ideal if you have a communal space you can use at school. That way, all members can have regular contact with the experiment. However, it is possible that the experiment will need to be set up at the home of one of your group members or at some other site. If you have a choice in this matter, choose a location that allows the most group members access to the experiment.

Assigning Tasks to Group Members

Second, it is important that all group members participate in data collection. At your first meeting, you should make a list of (a) the tasks that are involved in setting up the experiment and (b) the data collection tasks that will be required once the experiment has begun. Here is a general list to get you started, but your list should also include very specific tasks that are necessary for your experiment.

- Purchase or otherwise obtain laboratory equipment

- Organize the laboratory notebook (before the experiment begins)

- Measure and record quantitative data

- Write and record qualitative (descriptive) data

- Ensure that experimental procedures follow proposed methods

- Monitor possible influence of external variables

- Monitor safety and ethics issues

- Contact the teacher or mentor with questions (one person in the group should be assigned this task so that the teacher or mentor is not inundated by all group members with identical questions)

- Take photographs and label them in the laboratory notebook

Once you have a list of tasks, talk openly with group members about what you personally would like to do. The goal is to evenly distribute tasks, taking into account the strengths of each member. If one group member enjoys organization, assigning a task to this member, such as designing data tables for the laboratory notebook, will enable the group to be more efficent. If another member says that he or she is detail oriented, that person might be given the quantiative data collection tasks. Each member, however, will have multiple tasks and, therefore, *you will probably have a task that does not match what you are best at. Working in a group includes taking on tasks that stretch you as an individual.* Once your group agrees on which tasks each member will perform, you might consider assigning offical individual titles—for example, Notebook Manager, Quantitative Data Recorder, Descriptive Data Recorder, and Experimental Design Monitor.

As you assign roles to each group member, be sure that the distribution of tasks does not compromise your experimental design by adding external variables. For example, it would not be a good idea to rotate tasks among group members every two weeks in order to equalize the work load. Individuals might have slightly different techniques that could introduce variations into the experiment. It is better to assign specific tasks to one individual who performs those tasks throughout the entire experiment.

Once tasks and roles have been agreed on, write up a contract and have each group member sign it. Give a copy of the contract to your teacher so he or she can keep track of what each group member is supposed to be doing. Also, put a copy of the contract in the laboratory notebook so you can refer to it throughout the experiment.

Organizing a Group Laboratory Notebook

A third issue to address when working with a group is whether each individual will keep a separate notebook or if the group will keep one consolidated notebook. There are pros and cons to each.

Reasons for Keeping Individual Laboratory Notebooks (Paper or Online)

- Ensures that you will have to communicate with your group members

- Creates accountability on an individual basis.

- If one notebook is lost, other group members still have copies.

- Encourages individual members to record inferences that may differ from those of other group members.

However, keeping individual notebooks is not convenient if the experiment is located off site, and each member has various roles in the data collection.

Reasons for Keeping Group Laboratory Notebooks (Paper or Online)

- All data are in one location. Each group member is responsible for performing tasks and entering data as assigned to him or her.

- If the notebook is online, all members have access to it no matter where the experiment is being conducted.

- If the notebook is online, one member can post assigned tasks to the appropriate portion of the notebook and other group members can enter data *simultaneously.*

- Group notebooks foster collaboration among group members because each member can post comments and inferences as the experiment is conducted.

However, if one member of the group does not have as reliable online access as the others, it may be difficult to distribute the tasks evenly or to expect equal collaboration.

Components of a Laboratory Notebook

Although researchers organize their laboratory notebooks differently, there are certain basic components that are found in most notebooks (Beavon 2000;

Gordon 2007; Karnare 1985). For our purposes, your laboratory notebook will be subdivided into seven sections. The headings for these sections will be as follows: Title Page; Table of Contents; Experimental Proposal; Record of Procedures; Record of Correspondence; Data Tables; and Graphs, Statistical Analysis, and Conclusions. These sections are the suggested minimum for a laboratory notebook. With the approval of your teacher, you can add, remove, or rename section headings to better fit your experimental design.

Title Page

The Title Page is the first page of your laboratory notebook and should be dedicated to providing introductory information about its contents. Most important, write the title of the experiment along with the dates the experiment will be conducted. Below this, write your full name, address, contact information (either your e-mail address or a phone number), school name, and teacher name. This will help the notebook find its way back to you should it be misplaced. For online laboratory notebooks, the title page is simply the home page of your website. However, do *not* post personal contact information online and be sure to follow any additional restrictions your school has for posting identifiable personal information.

Table of Contents

The next two pages should be reserved for the Table of Contents. Here you list the section headings as well as the page numbers on which each section begins. Therefore, you will need to number every page of your laboratory notebook; do this when you first get the notebook. Use the seven section headings listed in this chapter (unless you have made other arrangements with your teacher). You may also need to add subsection headings. For online laboratory notebooks, write a paragraph on the home page/title page explaining how to navigate the online notebook. This paragraph will accomplish the same function as a table of contents.

Experimental Proposal

The next section, under the heading Experimental Proposal, should be a record of the design proposal your teacher approved. Therefore, this section will include a project title, an introduction (if it was required), an experimental design table, your hypothesis, materials, methods, and possibly an appendix.

Since the proposal was typed, you will need to reduce the font or page size so that you can directly attach each page of the proposal on a single notebook page (rubber cement or scrapbooking adhesives work best). Don't attach a stapled packet that needs to be unfolded to be read. You want to have quick and easy access to this document. For online laboratory notebooks, you may choose to copy and paste your proposal directly onto the Experimental Proposal web page, or you could provide a link to a Google Doc. Uploading the proposal as an attachment is not recommended because it is not as quickly accessible.

Record of Procedures

Next, your laboratory notebook will include a section with the heading Record of Procedures. Reserve a large section of blank pages to be used for this section. When you begin your experiment, you will refer often to the methods section you wrote for your proposal. In the Record of Procedures section, you record the date and a detailed description of what you completed each day of the experiment. It is critical to record detailed and accurate entries.

The pages of this section should be organized to look something like Table 6.1. Notice several aspects about this setup. The first column is labeled "Day of Experiment"; the second column is for the *calendar date*. The information in the first column is particularly helpful because, in your discussion of the results, you will be asked to refer to the experiment day, not specific dates.

Rows can remain blank if the experiment is not attended to daily. If you want to, use the pronoun *I* in the column headed Procedures Completed. If for some reason, you must be away on a day when you had intended to work on your project, have someone else collect data, take measurements or readings, and perform tasks to maintain the environment. If that happens, write down the name of the person who performed the tasks on those days (at the end of his or her entry). This information might prove to be significant later on in the research process. For online laboratory notebooks, the Record of Procedures could be directly on the web page or it could be a link to a Google Doc.

Accuracy and detail are extremely important when recording the procedures. They make you accountable to yourself and to your teacher as well. Your teacher will occasionally ask to see your laboratory notebook and will refer to this Record of Procedures section to be sure that you are in fact doing what you proposed to do and that you are writing down these procedures in sufficient detail.

Record of Correspondence

The Record of Correspondence section of your laboratory notebook is for documenting any correspondence or communication that relates to your experiment. Most likely the communication will be between you and your teacher or mentor. This documentation is important because it might include advice or suggestions for how to continue or modify your experimental design after

6

Table 6.1

Sample Record of Procedures

Day of Experiment	Date	Procedures Completed
1	Mon. Oct. 7	Set up control and experimental groups. (Notice that the day you set up the experiment is day 1).
2	Tues. Oct. 8	Watered control group and experimental group #3 until the water ran out the bottom. After 10 minutes I emptied the saucer.
3	Wed. Oct. 9	‑‑ ‑‑
4	Thurs. Oct. 10	Added more soil to experimental group #2 (Fluffy, my sister's cat, tipped it over!)
5	Fri. Oct. 11	‑‑ ‑‑
6	Sat. Oct. 12	Watered all groups—used same procedure as 10/8

the experiment has begun. Put summaries of these interactions in a list that is organized similar to the Record of Procedure table. For both written and face-to-face correspondence, record (a) the date of the interaction, (b) a detailed description of what you discussed, and (c) what conclusions you and your teacher or mentor reached. If you received permission to modify your proposed methods, be sure to provide an explanation of why this change was considered to be necessary.

If you are working with a mentor, you will probably have e-mail exchanges regarding your research. Classroom teachers may also provide feedback electronically. These written interactions should be printed out, dated, labeled, and directly attached into this section of your laboratory notebook. Therefore, save extra blank pages for these exchanges to be included in this section. You should still write summaries of the correspondence into the dated list, but then refer to the printed documents as shown in the second entry in Table 6.2 (p. 86).

If you are working with a group, it is important that any discussions that directly affect the experiment are recorded in the Record of Correspondence section. One entry might be if the group assigned or changed roles for each member; this should be documented as shown in the first entry of Table 6.2.

For online laboratory notebooks, the Record of Correspondence table could be directly on the web page (or in a Google Doc) and the Record of Correspondence could be an additional page or link. You could put copies of the e-mails (copied and pasted) right onto the additional pages. Remember to date and label each e-mail. Enter summaries into the table, refer to the e-mail, and then create a link directly to the e-mail copy.

Table 6.2

Record of Correspondence

Day of Experiment	Date	Procedures Completed
Prior to start of experiment	Thurs. Oct. 3	Group decided to divide the data collection workload. I (Jose) will be the laboratory notebook manager, in charge of setting up and maintaining the online laboratory notebook space, making sure all group members enter data as scheduled, monitoring the experiment for possible influence of external variables, and recording any correspondence. Sam will be the quantitative Data collector in charge of taking and recording all measurements into the data tables. Shawna will be the descriptive Data collector in charge of writing and entering descriptive data into the data tables. See signed contract for full details.
6	Sat. Oct. 12	When the experimental groups of bacteria cultures showed no sign of growth in the first 5 days of our experiment, I emailed our mentor (see email #8). Because the control plate is growing adequately we suspected that we had inoculated the plates correctly, but that maybe the concentrations of the experimental group were too strong. He then helped us recalculate concentrations so that we can inoculate new plates with the new concentrations early next week.

Data Tables

A large section of your laboratory notebook will be the Data Tables section. This is the space dedicated to recording data from your experiment. It is important that this section be well organized. (*Note:* Never jot down data in some random place with the intention of transcribing the data later into the notebook. You probably won't remember to enter that data.) Before you begin your experiment, you should spend some time thinking about the data you plan on collecting because the types of tables you decide to use are determined by the research design of your experiment. Refer to Tables 6.3 and 6.4, which suggest ways to organize your data in data tables.

Both raw data and descriptive data are best organized into tables or cells. The data that you collect during your experiment is referred to as *raw data* because no calculations or statistics have been applied to these numbers. These tables can be constructed in a word-processing program, printed out, then permanently attached within the Data Table section of your laboratory notebook. Or you can draw the tables by hand directly into the notebook.

If your laboratory notebook is online, you have several options. Word-processing files, as well as spreadsheet files, can be posted to Google Docs. Just be sure to organize these files all in one space and make sure that they are correctly linked. For example, if you organize your laboratory notebook in a wiki, you would have an entire page titled Data Tables. On the Data Tables page, you could have links to each of the data tables where data is being entered. Although you could design tables directly on wiki pages, it is ideal to link to word-processing or spreadsheet files posted in Google Docs because several individuals can be working in the document simultaneously. No matter how you choose to organize your data, be sure to design these tables specifically for *your* experiment.

Because you are collecting different types of data, you will most likely need to design various tables with different organizations. For example, you may be collecting several quantitative measurements, along with several descriptive observations. The example shown in Table 6.3 puts both quantitative and descriptive data together in the same table. Notice that the data is for just one entity. This type of data organization requires that each entity be recorded in its own table (rather than all entities being recorded in the same table).

The data table example shown in Table 6.4 (p. 88) describes only qualitative (descriptive) data. This means you would put the quantitative data in another table. This method of organization allows you to write out your inferences or your explanation of why you think the data are occurring the way they are. These inferences should be based on (a) what you are observing **and** (b) what you learned in your background research. Having an inference

Table 6.3

Sample Quantitative and Qualitative (Descriptive) Data Table

Day #	Date	Quantitative Data		Qualitative Description of the Plant, Including Environmental Change
		Leaf Length (mm)	Leaf Width (mm)	
1	Mon. 10/7	88	15	Leaf is dark green in the center, but edges are crispy and beginning to turn brown.
2	Tues. 10/8	89	15	Dark brown edges are more noticable and have grown toward the center of the leaf.

column may help you when you go to analyze your data because you have been recording your thinking all along during the experiment.

Table 6.5 is an example of a quantitative data table of someone who thinks ahead! Notice that the bottom row contains a place where total changes can be calculated after the data have all been collected. If measurements are entered directly into a spreadsheet program, you can set up the spreadsheet software to perform the calculations automatically.

Table 6.4

Sample Descriptive and Inference Data Table

Day #	Date	Qualitative Description	Inference (What This May Mean)
1	Mon. 10/7	Leaf is dark green in the center, but edges are crispy and beginning to turn brown.	This is the first change I've noticed. The edges may indicate that the over-watering is affecting the plant.
2	Tues. 10/8	Dark brown edges are more noticable and have grown toward the center of the leaf.	As more of the leaf turns brown, less of the surface area can be used for photosynthesis.

It is important to monitor and record the influence of external variables. This can be done by taking measurements and by writing descriptions. These measurements and descriptions should be recorded in a separate data table. For example, temperature, humidity, pH, evaporation rate, and respiratory rate, as well as other influences such as light or radiation, can influence the

Table 6.5

Sample Quantitative and Total Change Data Table

Day #	Date	Plant #1: Leaf Length	Plant #1: Leaf Width	Plant #2: Leaf Length	Plant #2: Leaf Width	Plant #3: Leaf Length	Plant #3: Leaf Width
1	Mon. 10/7	75	15	61	22	89	18
2	Tues. 10/8	77	15	61	22	89	18
10	Thurs. 10/16	82	18	59	25	93	18
Total changes		7 mm	3 mm	−2 mm	3 mm	5 mm	0 mm

Notice that when growth decreases rather than increases the number is negative.

When no change at all has occurred, a zero is recorded.

outcome of the experiment if they are not maintained for all groups within the experiment. External influences vary for each experimental project, but Table 6.6 shows an example of how data could be organized. This data table has columns for experimental day, date, specific measurements, and then a column for descriptive influences.

Table 6.6

Sample Influence of External Variables Data Table

Day #	Date	Room Temperature	Descriptive Influences
1	Mon. Oct. 7	22°C	— —
2	Tues. Oct. 8	19°C	The lights were off when I came in to take measurements today. This may be why the temperature of the room was lower than yesterday.
3	Wed. Oct. 9	21°C	— —

Documenting Evidence Using Photographs

In addition to writing descriptions and recording numbers in data tables, you should consider taking photographs of your experiment. Photographs are an excellent way to document things like the experimental setup and the entities being studied for data collection. Your teacher may even require you to add photographs to your final paper; they also make great additions to posters if your final project is going to be presented at a poster symposium or science fair. Consider taking the following photographs:

- You, the researcher, shown working on your experiment

- Experimental setup, to show the overall environment

- Individual photos of the experimental and control groups on the first and last day of the experiment (and maybe additional photos throughout)

- Close-ups of how data were collected—for example: a close-up photograph of your hands holding the instrument to take measurements.

Be sure that it is easy to determine important information about each photo. Consider labeling each entity with labels or index cards marked with the date or how long the experiment has progressed, such as shown in Figure 6.1. Be sure to write in dark permanent marker or a thick font so it will show up in the photographs. Or add these labels later using photo editing programs.

Figure 6.1

Photography Cards

Control Group pH 5.5 Day 7	Experimental Group #3 pH 6.0 Day 7

Consider these tips for taking good photos:

- If you put more than one entity in a photo, place each entity in the same order every time you photograph the entities. That will make it easier to discuss the differences when you begin analyzing the results.

- Determine the prime spot to take photographs. Take photographs with the same background every time. If you can use natural light instead of flash, do it.

- Be careful not to take photos facing straight into windows or mirrors or you will get a bad reflection.

- Get as close as you can to your entity when photographing it. Many cameras have a "macro" setting that allows you to get within centimeters of your subject and still remain in focus. The symbol is usually a tulip-shaped flower.

Photographs that capture the experimental setup and entities throughout the experiment belong in the laboratory notebook. Place setup photos in the Record of Procedure section; photos of the entities belong in the Data Table section. *Be sure to date, accurately label, and describe each photograph.*

Use Technology to Organize, Share, and Protect Photographs

There are many ways to use technology to help organize, share, and protect the photos you have taken throughout your experiment. Consider storing your photos online. Photography-sharing sites, such as Flickr, have annotating features, which allow you to highlight portions of the photos to tag. You could add quantitative measurements and/or descriptive notes about each

90

entity right "on" the photo. This leaves a box on the photograph, and when hovered over, a pop-up window displays the text.

But photography is not your only option. Consider making short videos during the research project. Record how you collected data, along with the results. Edit these videos into short digital stories and post the videos to a class or individual blog. Or post them to a video-sharing website such as You-Tube, TeacherTube, or Google Videos.

If your laboratory notebook is organized online, you should directly put the photographs in the appropriate sections or provide links to where they are. For example, on your Data Tables page, in addition to links to your online data tables, you could include links to Flickr. Each photograph must be dated, accurately labeled, and described.

Graphs, Statistical Analysis, and Conclusions

The last section heading of your laboratory notebook is Graphs, Statistical Analysis, and Conclusions. You will use this section after the experiment is complete. This section is the place for you to begin to statistically determine whether or not your hypothesis was supported. You will reorganize your raw data into tables and graphs to show comparisons and calculate statistical tests to determine significance of your results. All graphs, statistical tests, and calculations, whether completed longhand or using technology, should be inserted (clearly labeled) in this section. If you use technology, be sure to *print out results* and paste them in your notebook, even if you don't think you'll use this information in your final paper. Next to the calculations and graphs, record your thoughts about what they may indicate about your experiment. More specific instructions for analyzing your data and recording these ideas into the laboratory notebook are in Chapters 7, 8, and 9.

Conclusions

Keeping accurate, honest, and reliable records is critical to the research process. On the days that you are collecting data, you should be writing in more than one place in your laboratory notebook. For example, you are likely to be writing in

- Record of Procedures (to account for what you completed that day, maybe inserting photographs taken)
- Data Tables
  - Quantitative data tables for each entity
  - Qualitative (descriptive) and inference data table
  - Influence of external variables data table

Chapter Questions

1. Why is it important to keep an accurate laboratory notebook?

2. What are the pros and cons of paper laboratory notebooks? What are the pros and cons of online laboratory notebooks?

Chapter Applications

Decide whether you want to organize a paper or online laboratory notebook, and begin construction. The more time you put into organizing it now, before you begin your experiment, the less hassle you will have once you begin collecting data. Tailor the ideas in this chapter to fit your own experimental design. The sample data tables provided in this chapter are just that—samples. The primary goal is to organize your data so that they are ready for statistical analysis later. The ways you might choose to record the data are less important than the fact that they are organized and all in one place.

Once your teacher has given you the offical experimental start date, you may begin your experiment. The key now is to actually do what you have said you would do in your proposal. If you run into issues you did not anticipate, especially if you are considering varying from the methods you proposed, talk to your teacher. After you have received approval to make changes, record this approval in the Record of Correspondence section of your notebook and move forward. The next three chapters will help you as you begin to organize and analyze your data for analysis.

If working with others on your STEM research project, assign roles to each group member and write a contract agreement for all members to sign. Then turn the document in to your teacher (see also p. 80).

References

Beavon, R. 2000. Writing the laboratory notebook. Retrieved March 23, 2011, from *www.rod.beavon.clara.net/lab_book.htm*.

Gordon, J. C. 2007. *Planning research: A concise guide for the environmental and natural resource sciences*. New Haven, CT: Yale University Press.

Kanare, H. M. 1985. *Writing the laboratory notebook*. Washington, DC: American Chemical Society.

Purrington, C. 2009. Advice on keeping a laboratory notebook. Retrieved March 23, 2011, from Swarthmore College website: *www.swarthmore.edu/NatSci/cpurrin1/notebookadvice.htm*.

—7—

Descriptive Statistics

Introduction

Congratulations! You have completed your experimental data collection and are now ready to organize and analyze your data to determine whether or not your hypothesis is supported. Chapters 7, 8, and 9 of this *STEM Student Research Handbook* contain guidelines on how to graphically represent data and provide tips on using descriptive and inferential tests in STEM research.

Recording Calculations in Your Laboratory Notebook

I discussed in Chapter 6, "Organizing a Laboratory Notebook," how important it is to write down everything you do for your experiment in one place—in either a paper or an online laboratory notebook. By "everything," I mean the organization of your data and the application of statistical tests to your experimental research. In other words, in your laboratory notebook, don't put only the final polished tests you plan on using in your research paper or presentation but rather all the calculations you make to help you determine what is significant and what is not. Therefore, your laboratory notebook will be full of charts, graphs, tables, calculations, and computer printouts that may never make it into your paper or presentation. Remember, one of the main purposes of the laboratory notebook is to provide a place for you to record your observations, inferences, and analyses throughout the process. And recall the laboratory notebook guidelines from Chapter 6: never rip out pages (or delete text), and never use correction fluid to hide changes or errors. Simply cross out numbers with a single line.

When beginning statistical analysis in your laboratory notebook, you should clearly label which statistics you are using on what data. Write out—directly on the notebook page—preliminary tables to help you in your calculations. If you use a calculator to record the results, round only at the end

Key Terms

Arithmetic mean: A measure of central tendency that is the centermost point when there is a symmetric distribution of values in the data set; also referred to as an *average*.

Bimodal: A data set in which there are two clear data points that are represented more often than the other data.

Central tendency: A group of descriptive statistics that measure the middle or center of a quantitative data set, whose value, or number, best represents the entire data set.

Data set: The collection of similar data recorded within a researcher's laboratory notebook.

Descriptive statistics: Statistics that describe the most typical values and the variations that exist within a data set.

Interquartile range (IQR): A type of statistical variation that presents a data set as a graphical representation that indicates the range of data organized into four quartiles.

Median: A measure of central tendency that represents the number that appears in the middle of an ordered, finite list of data.

Mode: A measure of central tendency that represents the value that appears most often in a data set.

Outliers (outlier data): Data points that seem to lie outside the data set and do not appear to belong with that data set.

Range: A type of statistical variation that indicates the difference between the largest and smallest values in a data set. It is a measure of how spread out the data are, and therefore it is sometimes called the *spread*.

Standard deviation (σ, SD, s): A commonly used type of statistical variation that measures how close data are from the mean.

Statistical variation: A measure of how scores differ from one another; also known as *variation, statistical dispersion, spread,* or *dispersion*.

Unimodal: A data set in which there is one clear data point that is represented more often than the other data.

Variance (s^2): A type of statistical variation that is the standard deviation squared.

---

of the calculations by using the memory feature of the calculator. It is important to correctly apply the order of operations and put every number with the right expression in the technology you use. If you are using Microsoft Excel or another statistical program, use printouts as a record that you performed the statistical test. Always write down your interpretation of the statistics and thoughts regarding the usefulness of the statistical test in relationship to what it tells you about the data within your experiment. Use formal, scientific language. An *unacceptable* laboratory notebook entry would be:

Cool, finally after messing with those gross bean sprouts, the t-test *shows that the number of plants makes a difference to how many of these weird things actually come up.*

An *acceptable* laboratory notebook entry should be an interpretation of the statistics including explanations about how results might introduce new questions. For example:

The t-test *results support the hypothesis, and therefore the number of bean sprouts per square cm of soil does have an impact on speed of germination. I wonder if, based on this, I may be able to determine the optimal number of seeds to plant in order to shorten germination time of an entire crop.*

If you are using a paper laboratory notebook, either write directly within the notebook or attach printouts to the page with your personal interpretation nearby. If you are using an online notebook, the same principles apply except that you will have links to Excel (or other statistical software) files showing the same processes. When performing statistical tests using computer technology, refrain from running statistics and simply deleting the results when they do not appear to highlight important findings. The laboratory notebook is a record of what works *and* what does not. Therefore, use the "worksheet" function in Excel to record the many analyses that you will perform on your data. Label graphs and mathematical calculations clearly, and write down your analysis for each descriptive statistic, graphical representation, or statistical test before moving on to another.

Introduction to Descriptive Statistics

Descriptive statistics are statistics that describe the most typical values and the variations that exist within a data set (Salkind 2008). The term *data set* refers to the numerical data you recorded as the results from your experiment. For example, a data set might include the measurements you recorded for the 20 trials of your experimental and control groups. The most common way to describe data is by using the measures of central tendency and the statistical variation.

Measures of Central Tendency

The measure of central tendency is the one value of a quantitative data set that is most typical (Cothron, Giese, and Rezba 2006). This number is used to best represent the entire data set. There are three types of central tendency that will be discussed in this section: *mode, arithmetic mean (average),* and *median.*

Each represents the entire set of values but highlights the central tendency of a distribution differently. Measures of central tendency can be used in isolation to analyze your data; however, being able to calculate mode, mean, and median are critical to performing inferential statistics that are introduced in Chapter 9.

Technologies for Calculating Statistics

Calculators
Texas Instruments *http://education.ti.com/educationportal/sites/US/homePage/index.html*
Casio *www.casio.com/products*

Spreadsheet Software
Open Office *www.openoffice.org*
Microsoft Excel *http://office.microsoft.com/en-us/excel*
Spreadsheet function in Google docs *https://docs.google.com*

Statistical Software
Key Press Fathom *http://keypress.com*
SPSS *www.spss.com*
Minitab *www.minitab.com*
PSPP (open source) *www.gnu.org/software/pspp*

Online Statistical Tutorials
Stat Tutorials *www.stattutorials.com*
Stat Tutorials for performing statistics in Excel *www.stattutorials.com/EXCEL/index.html*
Elementary Statistics and Probability Tutorials and Problems *www.analyzemath.com/statistics.html*

Mode

The *mode* is the value that appears most often in a data set (Salkind 2008). For example, your teacher may use this central tendency to know which of the items on a test was the most difficult for the class—it will be the item most students missed. For another example, suppose you collected temperatures for 11 days. You recorded the following temperatures in Celsius:

12, 12, 13, 14, 14, 15, 15, 15, 15, 37, 39

Both 12°C and 14°C appeared twice and 15°C appeared four times. Since 15°C appeared the most, this is called the mode of the data. The term *unimodal* means that there is one clear data point that is represented more often then the other data. The mode is not necessarily unique because two values can have the same frequency in the same data set. One of the disadvantages of using mode is that many data sets have more than one value that is represented. For instance, in the data set {2, 3, 3, 4, 5, 5} both the number 3 and the number 5 are represented twice; therefore, this data set has two modes, 3 and 5 (which is referred to as *bimodal*). In addition, if the distribution is uniform, as in this data set {5, 5, 5, 5, 5, the mode is 5 but has little meaning and should not be used. The mode is best used with other measures of central tendency but not alone.

The mode should be used (1) when the data are categorical in nature or (2) when the data are not uniform. For example, if we want to know which tools and instruments are most used in a chemistry laboratory, each item would be tallied into a category (e.g., thermometers, beakers); the item with the most tallies would be the mode and would represent the most used items. Mode would also be appropriate to use in an experiment that has kilogram measurements of {65, 68, 69, 71, 72, 73, 73, 75, 77} because the data are not uniform and vary within the set. However, if the data set has no mode {65, 68, 69, 71, 72, 73, 75, 77}, then it should not be used as a measure of central tendency.

Arithmetic Mean (Average)

The arithmetic mean is commonly called an *average* or a *mean*. The mean is a measure of central tendency that is the centermost point when there is a symmetric distribution of values in the set (Weiss 2008). In other words, it is the centermost point because all the values on one side of the mean are equal in weight to all the values on the other side of the mean. Therefore, the arithmetic mean is very sensitive to extreme values, which will make it less representative of the set of values and less useful as a measure of central tendency for data sets with extreme values.

To calculate the mean, you add all the values and divide by the total number of values. The mathematical formula for mean is as follows:

$$\text{mean} = \frac{\sum x}{n}$$

where $\sum x$ represents the sum of all the values within the data set and n represents the total number of values within the data set.

If you had multiple entities or trials within a group, you may want to calculate a mean so that the data in each group are combined for comparison. For example, if the trial measurements for density of a control group were 8 g/cm^3, 10 g/cm^3, 7 g/cm^3, and 9 g/cm^3, the arithmetic mean would be calculated by adding all the measurements (or data points) together and dividing by the total number of measurements (or data points) of the entire group.

$$\frac{8 \text{ g/cm}^3 + 10 \text{ g/cm}^3 + 7 \text{ g/cm}^3 + 9 \text{ g/cm}^3}{4} = \frac{34 \text{ g/cm}^3}{4} = 8.5 \text{ g/cm}^3$$

Means are a good statistic to use if your data are unimodal and approximately symmetrical on each side. For example, in the following ordered list, the single data point that is represented more often than any other number is the number 51.

12, 20, 33, 42, 48, 49, 50, 51, 51, 51, 56, 58, 64, 77, 83, 90

Having symmetrical data means that the data are divided in half—that the two sides are identical. Therefore, the numbers above are unimodal and symmetric.

If your data are not unimodal or symmetric, calculating means could produce results that are misleading. For example, suppose you collected temperatures for 11 days. You recorded the following temperatures in Celsius:

12, 12, 13, 14, 14, 15, 15, 15, 15, 37, 39

The mean, or average, is calculated as follows:

$$\frac{12 + 12 + 13 + 14 + 14 + 15 + 15 + 15 + 15 + 37 + 39}{11} = \frac{201}{11} = 18.27°C$$

What temperature was most common in this data set? Was it 18°C? No! Clearly from the data above, the most common temperatures were in the early to midteens. If you were to look at a histogram or dot plot of this data, you could see that it is bimodal. That is, there are two obvious peaks of the data, one at 15 and the other at 39. Therefore, the mean, in this case, is *not* the best measurement to use to calculate and display your data. Because of the extreme value, calculating the median would be a more accurate way to represent this data.

Median

The *median* is the number that appears in the middle of an *ordered* finite list of data (Cothron, Giese, and Rezba 2006). A median is the number that separates the higher half of the data set from the lower half. Median can be found by ordering the list from lowest value to highest value and then choosing the middle number. Therefore, the median of an odd-numbered data set is the center value; but for an even-numbered data set the median is the calculation of the mean of the two middle values. The median should be used when the data set has extreme values. Unlike the arithmetic mean, the median is insensitive to extreme values. The median is the centermost value of the data set even when the data set has one or more extreme values.

The difference between mean and median is that the arithmetic mean is the middle point of the set of values, and the median is the middle point of the number of data in the data set. Therefore, the median is an indication about how many data there are in the data set, not the values of those cases. In the data set below, there are 11 values. The middle number is the sixth number because there would be five data points on each side. Therefore, the median of this data set is 15.

12, 12, 13, 14, 14, 15, 15, 15, 15, 37, 39

Comparing the Three Types of Central Tendency

We have looked at three different types of central tendency: mode, arithmetic mean, and median. In a set of data with normal distribution, all three are the same value. However, when data are not distributed normally, the values of central tendency will vary. The mean is used with many inferential statistical tests like the *t*-test and ANOVA. (See Chapter 9 for more information about inferential statistics.) The mean is not the centermost value of a data set when the data set has extreme scores, in which case you would use the median. However, when the data set is categorical, mode should also be used for central tendency. Remember that the mode should not be used as the only measure of central tendency; it should always be used with another measure of central tendency.

Statistical Variation

Another way to describe the data is by using statistical variation. A measure of *statistical variation* reflects how scores differ from one another (Salkind 2008). These differences are also called *variation, statistical dispersion, spread,* or *dispersion.* This measure indicates how different scores are from one another. Even if two sets of data have the same mean, the statistical variation can be different and describe how different the data sets are. Variation represents the average difference from the mean. The variation increases as the data become more diverse. Below, I discuss the four measures of variation: *range, interquartile range, standard deviation,* and *variance.*

Range

Range is the most general measure of variation. It is a type of statistical dispersion that indicates the difference between the lowest and highest values in a data set (Triola 2001). It is a measure of how a set of measurements or data is spread out and is sometimes called the *spread of the data.* In other words, the range gives an idea of how far apart values are from the lowest to the highest. Range is calculated by taking the difference between the highest and the lowest number in the data set.

For example, in the temperature data on page 98, the range would be the highest temperature measured minus the lowest temperature measured. The highest temperature measured is 39°C and the lowest temperature measured is 12°C, then the range is 39°C – 12°C = 27°C. Therefore, the range, or spread, of this data is 27°C.

Interquartile Range

Another measure of statistical dispersion is interquartile range (IQR), also called the *midspread* or *middle fifty*. IQR is a graphical representation of a data set that indicates the range of the data organized into four quartiles (Cothron, Giese, and Rezba 2006). To calculate the IQR, you will need to determine the upper and lower quartiles. These numbers will separate your data into four quartiles, or quarters, each containing 25% of the data. To do this, find the median of the entire data set. Next, identify the median of the lower half of the data. This is the lower quartile median. For the upper quartile, find the median of the upper half of the data.

The median of the data set in Figure 7.1 is 15°C. The lower quartile median is 13°C and is labeled with the notation, Q1. The upper quartile is 15°C, and is labeled Q3. To calculate the interquartile range, subtract the upper and lower quartiles, 15 – 13 = 2°C. Therefore, IQR = 2. Calculating the interquartile range is very helpful for determining outlier data, which I discuss next, and for constructing a box and whisker plot (which I discuss in Chapter 8).

Figure 7.1

Finding the Interquartile Range

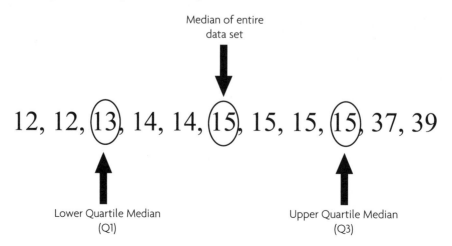

DESCRIPTIVE STATISTICS

Dealing With Outlier Data

Outlier data, or *outliers,* are data points that lie outside the data set and appear not to belong with that data set (Salkind 2008). Outlier data can occur within experiments due to either observational or recording error. If this is your first time performing the particular methods in your experiment, you can expect to make mistakes in the data collection process. Even professional scientists make errors when performing new methods. This is normal. If you confidently know that the outlier data is due to your own error, instrument error, or some other known cause *not* associated with your independent variable, you may consider throwing out outlier data when performing your calculations for data analysis.

Of course, these outliers must still be mentioned in the Results and explained in the Analysis and Conclusions sections of your paper or poster. Because outliers can taint the data, it is better to do calculations without them, but you must be confident that the outlier data is *not* related to the independent variable.

Consider the temperature data again. Suppose you noticed after recording temperatures of 37°C and 39°C that the thermometer had been broken. Therefore, instrument error is the known cause of the outlier data, and the two inaccurate readings can be left out of the calculations. If calculations are done only on the data set {12, 12, 13, 14, 14, 15, 15, 15, 15}, the results will not only be different but will show a more accurate picture of the scenario. On the other hand, if the two values 37°C and 39°C represent actual temperature measurements, they should remain in the data set.

How can you determine whether an outlier is a piece of data that does or does not belong? Mathematically speaking, there is an outlier "rule of thumb" to follow. Let y be a piece of data; y is an outlier if

$$y < Q1 - (1.5 \times IQR)$$
$$\text{OR}$$
$$y > Q3 + (1.5 \times IQR)$$

For example, in our temperature data:

$$Q1 = 13$$
$$Q3 = 15$$
$$IQR = 2$$

To determine whether or not 39 is an outlier, we would use the second equation with the Q3 number because our "outlier-in-question" is on the high end of our data set, not the lower end. Therefore

$$15 + (1.5 \times 2) = 18$$

Since 39 > 18, we have mathematical confidence that 39 is an outlier (as is 37). Use this calculation to help you determine which data you can confidently throw out when performing your statistical analysis. But remember, outliers still must be mentioned and addressed in your final paper or poster.

Standard Deviation

Standard deviation (SD) is the most frequently used measure of variation. SD is a measure of how close data are to the mean (Weiss 2008). It shows how much variation or dispersion there is from the arithmetic mean and represents the average amount of variation in a set of scores. The larger the standard deviation, the larger the average distance each data point is from the mean of the distribution. Therefore a low standard deviation indicates that the data points tend to be very close to the mean. If, on the other hand, the standard deviation is high, the data is spread out over a large range of values from the mean; scores can be close and still far away from the mean.

There are two types of standard deviation: one is calculated for a population and the other is calculated for a sample. As first introduced in the Chapter 2, "Research Design," *population* means the complete collection of every item that has the same characteristics. For your purposes, the population is the entire group from which you want to collect data. The *sample* is the subset of the population from which you actually collect data. Therefore, the sample is meant to be representative of the population. Statistically, all tests are more powerful with larger sample sizes (Triola 2001). For projects you design, you should have between 4 (the minimum) and 10 entities within each of your experimental and control groups. If your sample has more than 30 data points, additional, more powerful statistical tests can be used (Gonzalez-Espada 2007).

Getting information from the entire population is impractical and usually impossible in STEM research. Most likely, your research data represents a sample coming from a larger population. On the following page note the mathematical differences in the calculations of the standard deviation of a population compared to the standard deviation of a sample.

There are two important differences between the population and sample equations in the box. One of the differences is their symbolic notation: the standard deviation notation for *sample* is s and for *population* is σ. The other differences are the denominators of the fraction in the square root in each standard deviation. The standard deviation for a population divides by n and the standard deviation for a sample divides by $n-1$.

The standard deviation for a sample, which divides by $n-1$, is an unbiased estimate. By subtracting 1 in the denominator, the number is smaller and therefore the approximation of standard deviation is larger. Therefore, in case

Mathematical Notation for Standard Deviation in a *Population*

$$\sigma = \sqrt{\frac{\sum (x_i - \mu)^2}{n}}, \text{ where}$$

σ is the symbolic notation for standard deviation,
$\sum$ is sigma, which tells you to find the sum of what follows,
x_i is each individual value in the data set,
μ is the arithmetic mean of all the values, and
n is the number of values the data set has.

Note that $(x_i - \mu)$ tells you to find the differences between each individual value and the arithmetic mean.

Mathematical Notation for Standard Deviation in a *Sample*

$$s = \sqrt{\frac{\sum (x_i - \bar{x})^2}{n-1}}, \text{ where}$$

s is the symbolic notation for standard deviation (another symbolic notation is SD),
$\sum$ is sigma, which tells you to find the sum of what follows,
x_i is each individual value in the data set,
$\bar{x}$ is the arithmetic mean of all the values, and
n is the number of values the data set has.

Note that $(x_i - \bar{x})$ tells you to find the differences between each individual value and the arithmetic mean.

there is an error, the standard deviation will be an overestimation and will compensate for the possibility of errors, and therefore the standard deviation for a sample equation is a good choice for STEM research.

An example of how standard deviation can be calculated for an experiment is recorded in an article by Shaefer et al. (2000). The authors reported that Hurricane Hugo had a significant impact on stream water chemistry on tropical streams located in the El Yunque National Forest (which is located in Puerto Rico and is part of the U.S. Forest Service). Table 7.1 (p. 106) shows a sample of 10 randomly collected ammonia measurements (kg/hectare per year) in the first year after Hurricane Hugo (the hurricane occurred in September 1989).

Table 7.1

Ten Ammonia Measurements Taken in the El Yunque National Forest in the First Year After Hurricane Hugo

57	66	88	96	116
147	147	154	154	175

Find the arithmetic mean:

$$\bar{x} = \frac{57 + 66 + 88 + 96 + 116 + 147 + 147 + 154 + 154 + 175}{10} = \frac{1,200}{10} = 120$$

x	$x - \bar{x}$	$(x - \bar{x})^2$
57	$(57 - 120) = -63$	$(-63)^2 = 3969$
66	$(66 - 120) = -54$	$(-54)^2 = 2916$
88	$(88 - 120) = -32$	$(-32)^2 = 1024$
96	$(96 - 120) = -24$	$(-24)^2 = 576$
116	$(116 - 120) = -4$	$(-4)^2 = 16$
147	$(147 - 120) = 27$	$(27)^2 = 729$
147	$(147 - 120) = 27$	$(27)^2 = 729$
154	$(154 - 120) = 34$	$(34)^2 = 1156$
154	$(154 - 120) = 34$	$(34)^2 = 1156$
175	$(175 - 120) = 55$	$(55)^2 = 3025$
Total = 1,200		Total = 15,296

Then the standard deviation is

$$s = \sqrt{\frac{\sum (x_i - \bar{x})^2}{n-1}} = \sqrt{\frac{15296}{10-1}} = \sqrt{\frac{15296}{9}} \cong 41.23 \quad$$ kg/hectare per year. This indicates that there is an average distance from the arithmetic mean to the data of 41.23 kg/hectare per year. A standard deviation curve can be used to further explain the data.

As long as the distribution of data in the population is normal (which can be determined by a bell-shaped histogram), the addition of the mean and the standard deviation will provide additional insight into the data. The mean is represented in the highest peak of a normal standard deviation curve and is labeled as zero, as shown in Figure 7.2. Most of the data are found within 1 SD of the mean (34.13% on either side).

Figure 7.2

Standard Deviation Curve

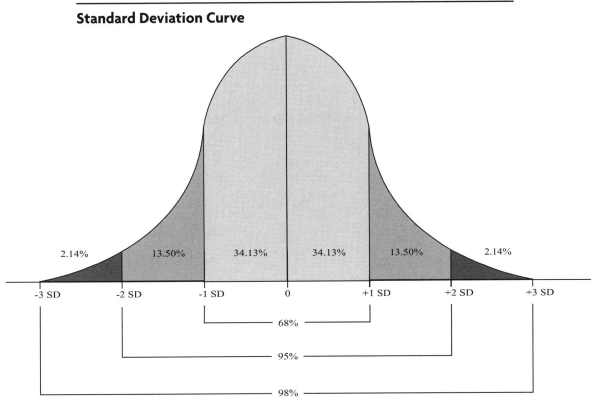

In the previous stream water chemistry example, ranges can be calculated by subtracting one standard deviation from the mean to show where 34% of the data are found. Since the mean is 120, we add or subtract 41.23 from each side to find the range of the first standard deviation.

$$120 + 41.23 = 161.23$$
$$120 - 41.23 = 78.77$$

These calculations indicate that the 34% of the data is in the range of 120 and 161.23 kg/hectare per year and that 68% of the data set is between the values 78.77 and 161.23 kg/hectare per year. Notice how the standard distribution helps describe the data. With the standard deviation, you can answer a lot of questions of your data, for example:

- What values were above the mean? What values were below the mean?

- What is the percentage of the data that were in a range of values? What 50% of the values are above the mean?

- What range of values would you expect for 95%?

The standard deviation is the average distance from the arithmetic mean (do not use median or mode). If every number is the same, the standard deviation is 0. Therefore, the larger the standard deviation, the more spread out the values are and the more different they are from one another. The standard deviation, like the mean, is sensitive to extreme scores.

Variance

The *variance* is the standard deviation squared (Weiss 2008). Once you have calculated standard deviation, you can just square it to get the variance.

The Mathematical Notation for Variance

$$s^2 = \frac{\sum (x_i - \overline{x})^2}{n-1}$$

s^2 is the symbolic notation for variance,
$\sum$ is sigma, which tells you to find the sum of what follows,
x_i is each individual value in the data set,
$\overline{x}$ is the arithmetic mean of all the values, and
n is the number of values the data set has.

Note that $(x_i - \overline{x})$ tells you to find the differences between each individual value and the arithmetic mean.

Notice that the variance and standard deviation are closely related and almost the same formula. The variance should always be reported with the standard deviation because the variance informs how far a set of numbers are spread out from one another in addition to the average distance from the arithmetic mean.

Standard deviation and variance can be calculated using spreadsheet software like Excel, or free online calculators such as *www.easycalculation.com/statistics/standard-deviation.php*.

Additional Calculations

There are additional calculations that can be done with your data, including total change and rate of change. If these calculations apply to your data, you should run a statistical test (like the *t*-test or ANOVA) to compare each experimental group's change to the control group in order to determine if the change is significant or not. See Chapter 9 for more information on inferential statistics.

Total Change

Total change measurements are easy to calculate, either by hand or in spreadsheet software such as Excel. These measurements help you compare the results of your experimental groups to one another as well as to the control group. The total change is calculated by taking the final measurement and subtracting the starting measurement.

$$\text{Final measurement} - \text{starting measurement}$$

For example, if day 1 had a measurement of 45 mm and the last day's measurement was 52 mm, the total change is calculated:

$$52 \text{ mm} - 45 \text{ mm} = 7 \text{ mm}$$

If total change increased in this group, the number will be positive, and if the measurement decreases, the number will be negative. If some entity's change varied (increased and decreased) throughout the experiment, total change may not be the most accurate description of your data.

If you had multiple entities or trials in an experimental group, you may want to calculate total change for each entity/trial, as well as an average total change for the entire group. For example, if beginning pressure measurements for an experimental group was 80 Pascals (Pa), 77 Pa, 85 Pa, and 76 Pa, and the ending measurements were 60 Pa, 77 Pa, 63 Pa, and 68 Pa, the average total change is calculated:

$$\frac{(60 + 77 + 63 + 68)}{4} - \frac{(80 + 77 + 85 + 76)}{4} = 67 - 80 = \text{-13 Pa}$$

Rate of Change

Rate of change is useful when you want to compare the speed at which changes occurred in a specific period of time. This is calculated by dividing the total change in this period of time by how much time is in this period.

$$\frac{\text{Final measurement} - \text{starting measurement}}{\text{Total time}}$$

For example, if an entity's weight changed from 7.2 kg to 6.1 kg in a span of 26 days, the rate of change is calculated:

$$\frac{7.2 - 6.1 \text{ kg}}{26 \text{ days}} = 0.042 \text{ kg/day}$$

Notice that the units used in the numerator and the units in the denominator remain in the final answer to indicate how much change occurred per unit of time. This is helpful when comparing groups to one another.

Using Descriptive Statistics to Explain Experimental Results

The purpose of calculating descriptive statistics, such as central tendency and variation, is to highlight specific characteristics of your data. Descriptive statistics will help you determine and discuss what differences, if any, exist between your experimental and control groups. Organizing these results into tables is a good way to present the data for the Data and Results section of your paper or presentation. For example, a table could contain the mean for each group as well as the standard deviation or variance. Ask yourself questions about the calculations, such as those listed in the "Data Interpretation" section of Chapter 9 on pages 140–142. Write down your observations and analysis about these descriptive statistics in your laboratory notebook.

However, any differences highlighted by descriptive statistics may not necessarily be a direct result of the changes you made in the independent variable (Valiela 2001). It is possible that these changes are due to experimental error or to chance. For example, if you determine a difference in the mean between the control group and one of the experimental groups, at this point you do not know if the difference is due to the random nature of data collection or if the differences are due to the treatment of the independent variable. This is why additional statistics called *inferential statistics* (see Chapter 9) may be necessary. The descriptive statistics that you calculated in this chapter will come in handy should you eventually decide to use inferential statistics to analyze your data. The next chapter, Chapter 8, is about describing your data visually using graphical representations. Organizing your raw data and descriptive statistics into graphs and tables will further help you determine what your data mean. Then, Chapter 9 will introduce you to the basics of inferential statistics—or how to determine whether your results are statistically significant.

Chapter Applications

Once you have collected data from your experiment, you will need to calculate various descriptive statistics explained in this chapter. Do this in your laboratory notebook, making notes to yourself about what the data may be saying about your results.

References

Cothron, J. H., R. N. Giese, and R. J. Rezba. 2006. *Science experiments and projects for students: Student version of students and research.* Dubuque, IA: Kendall/Hunt.

Gonzalez-Espada, W. 2007. Using simple statistics to ensure science-fair success. *Science Scope* 8 (30): 48–50.

Salkind, N. J. 2008. *Statistics for people who think they hate statistics.* Los Angeles: Sage Publications.

Shaefer, D., W. H. McDowell, F. N. Scatena, and C. E. Asbury. 2000. Effects of hurricane disturbance on stream water concentrations and fluxes in eight watersheds of the Luquillo Experimental Forest, Puerto Rico. *Journal of Tropical Ecology* 16 (2): 189–207.

Triola, M. F. 2001. *Elementary statistics.* New York: Addison-Wesley.

Valiela, T. 2001. *Doing Science: Design, analysis, and communication of scientific research.* New York: Oxford University Press.

Weiss, N. A. 2008. *Elementary statistics.* San Francisco: Pearson Addison-Wesley.

8

Graphical Representations

Introduction

Graphical representations are visual ways to represent raw data, descriptive statistics, or inferential statistics in order to highlight important findings from a research study. There are many different kinds of graphical representations; it is up to you to decide which best describes your data (Shaughnessy, Chance, and Kranendonk 2009). Then, you can use the graphical representations to determine additional statistical analysis that can be done to explain the significance of the data.

Learning Objectives

During the course of this chapter you should

1. familiarize yourself with the various graphical representations for quantitative and qualitative data,

2. determine which types of graphical representations should be used for your own data,

3. construct a variety of graphical representations in your laboratory notebook, and

4. make notes in your laboratory notebook about what each graphical representation highlights in your data.

Key Terms

Bar graph: Shows the distribution of qualitative data using bars that indicate frequency of a specific category.

Box and whisker plot: A graphical representation showing the distribution of quantitative data; the plot displays minimums, maximums, outliers, ranges, interquartile ranges, medians, and lower and upper quartiles.

Cartesian coordinate system: A system that plots points uniquely in a plane by a pair of numerical coordinates; the system indicates the distances from the point to two perpendicular lines (y-axis and x-axis).

Dot plot: A graphical representation showing the distribution of quantitative data using dots to show the relationship between two (or more) variables.

Frequency distribution: Shows the amount of variation within a spread of qualitative data that counts the frequency of data points of the values in the sample.

Graphical representation: A visual way to represent raw data, descriptive statistics, or inferential statistics in order to highlight important findings from a research study.

Histogram: A graphical representation showing the distribution of quantitative data using touching bins (see p. 115) that indicate frequency of a specific valve.

Inferential statistics: Mathematical calculations performed to determine whether the differences between two groups are due to chance or are a result of the treatment.

Line graph: A type of graphical representation using the Cartesian coordinate system that displays data points connected by line segments; often used to show change in time, with time being on the x-axis.

Normal distribution: A curved shape in a graph that shows increased frequencies centered around a single mean and fewer frequencies on either side of the mean.

Pie chart: A circular chart that represents data by dividing the whole circle into sectors by sizes that are proportional to the quantity the sector represents.

Scatter plot: A type of graphical representation using the Cartesian coordinate system that displays both the independent variable and dependent variable as a single point; often used to determine positive correlation, negative correlation, or no correlation between two variables.

Stem and leaf plot: A graphical representation showing the distribution of quantitative data to visually highlight the frequency of certain values, without losing the original, or raw, data.

Table: A type of graphical representation that organizes numerical and/or descriptive data in rows and columns that aid in the understanding of a large amount of data.

Construction of graphical representations can be done manually or by using technology such as a calculator or a spreadsheet program; consider a graphing calculator such as the TI-Nspire (or others in the TI family) or statistics software such as Fathom Dynamic Data Software.

Whatever method of construction you choose, you should enter all graphical representations into your laboratory notebook under the laboratory heading Graphs, Statistical Analysis, and Conclusions. As first discussed in Chapter 6, "Organizing a Laboratory Notebook," you should adhere, or link, to each table, graph, or statistical test and write down your observations and an analysis of what each graphical representation might show about the data. It is critically important that you label all graphical representations carefully so that you have an accurate record of exactly what data have been organized. By doing that, you will be able to accurately interpret the results in the days or even weeks after the data have been placed in your laboratory notebook.

Graphical Representations for Quantitative Data

The type of graphical representation you choose will be based on the type of data you have: quantitative or qualitative. This section introduces common types of graphical representations for displaying quantitative data: histograms, dot plots, stem and leaf plots, box and whisker plots, line graphs, scatter plots, and tables.

Histograms

A histogram is a graphical representation showing the distribution of data (McDonald 2009). At first glance, histograms may look just like what you have always called bar graphs. Although both types of graphs are constructed with horizontal or vertical bars, a histogram's bars are called *bins,* and these bins touch, with no spaces between them. In addition, the numerical data represented by the axis at the base of the bins are on a numeric continuum, with frequency being indicated by the height of each bin. Also, histograms are best used for quantitative data; bar graphs are better used for qualitative data. For example, a histogram should be used if you need to construct a

---

Tips for Generating Graphical Representations ("Visuals")

- Understand what it is you want to display

- Plan your visual before creating it

- Choose the best technology for the type of visual

- Communicate only one idea in a single visual

- Label everything to increase understanding

- Keep it simple; do not display too much information

- Maintain the scale by visually keeping equal distance between values on the *x*-axis and *y*-axis.

graph comparing the number of M&Ms in 23 different bags. But a bar graph would be used to compare the number of different color candies found in the bags of M&Ms.

Let's explore histograms further. Suppose the number of M&Ms in 23 separate bags are counted. Listed below are the raw data results with the amount of M&Ms in each bag of candy.

22, 23, 23, 23, 22, 24, 25, 21, 23, 24, 24, 23, 22, 21, 23, 23, 20, 24, 26, 22, 23, 22, 22

The numbers listed this way make it difficult to determine how many M&Ms are most commonly found, or least likely to be found, in a bag of M&Ms. Therefore, the first task is to arrange the data in order from least to greatest.

20, 21, 21, 22, 22, 22, 22, 22, 22, 23, 23, 23, 23, 23, 23, 23, 23, 24, 24, 24, 24, 25, 26

These data are perfect for a histogram. If we make it using vertical bins, our horizontal axis will be labeled with the varying quantities of M&Ms. Our data is best displayed by constructing bins that are one unit wide, starting with 20 and going up to 26 (see Figure 8.1). The vertical axis will be labeled *frequency,* or the number of bags that contained that quantity. Therefore, frequency represents the numbers that fall within the interval of that bin. There is only one bag that had 20 pieces of candy, so the first bin is one high. There are two bags that had 21 pieces so the next bin is two high, and so forth.

Figure 8.1

Histogram Frequency of M&Ms Found in 23 Bags

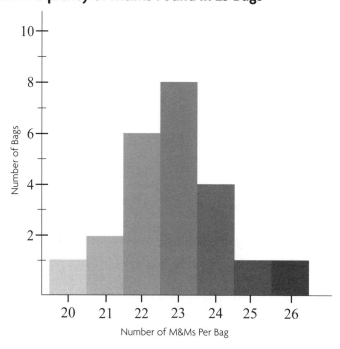

8

Figure 8.1 is an example of a *unimodal* histogram. This means there is one clear mode, or peak, in the data. This histogram can also be called roughly symmetric because both sides of the histogram look similar. Since this distribution has the largest bins at 22 and 23, it seems that 22 and 23 are the most common quantities of M&Ms that will appear in the bag. The bins are the smallest at the tails, or ends, of the histogram. It appears as if 20 and 26 are uncommon quantities of M&Ms. This curved shape that shows increased frequencies centered around a single mean and fewer frequencies on either side is known as a *normal distribution*. Because this sample of M&Ms has no bin that goes outside this curve, we can say the sample is normally distributed.

Constructing a histogram by hand is time-consuming and difficult to do well. To obtain professional quality, more efficient histograms can be constructed using either Texas Instruments software or calculators. The TI-84 and TI-Nspire families as well as Fathom statistical software can construct high-quality histograms.

Dot Plots

A dot plot, also called a strip chart or strip plot, is another way to graphically represent the distribution of quantitative data, using dots instead of bins. Similar to the histogram, dot plots have one axis labeled like a number line and another axis labeled with frequency (the quantity of data that falls within a given interval). When a dot plot is constructed, it will look similar to a histogram. In fact, the overall shape of the dot plot will be exactly the same as if you constructed a histogram. For example, consider the histogram of the M&Ms we just completed. Instead of making bins for each interval, we can place a dot for each M&M bag that falls into this category, as shown in Figure 8.2. Data within both dot plots and histograms can be oriented either vertically or horizontally. The dot plot in Figure 8.2 is shown horizontally. A dot plot can be constructed with the TI-Nspire and with Fathom statistics software.

Figure 8.2

Dot Plot Frequency of M&Ms Found in 23 Bags

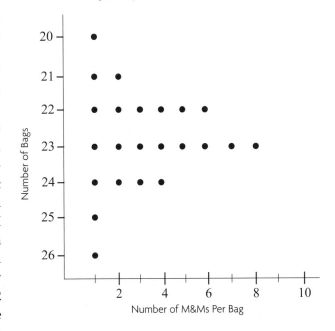

Stem and Leaf Plots

A stem and leaf plot is another way to display quantitative data. It is best used when you want to visually highlight the frequency of certain values, without losing the original, or raw, data. The distribution of this data will look similar to both a histogram and a dot plot, except the stem and leaf plot shows the numeric value of *each individual piece* of data in addition to grouping them as a whole. To organize your data, you will separate your data into both a stem and a leaf. The first column is called the stem and the second column is the leaf. The list of 20 values below represents weights in grams.

46, 34, 59, 44, 45, 55, 38, 47, 55, 31, 44, 56, 33, 45, 30, 58, 46, 49, 40, 48

Although not absolutely necessary, ordering the list makes it easier to plot.

30, 31, 33, 34, 38, 40, 44, 44, 45, 45, 46, 46, 47, 48, 49, 55, 55, 56, 58, 59

The first column, or stem, for our example will contain the tens digits while the second column, or leaves, will contain the ones digits. Therefore, the stem-leaf plot would look like Table 8.1.

Table 8.1

Stem and Leaf Plot of Weight in Grams

Stem (g)	Leaf (g)
3	0 1 3 4 8
4	0 4 4 5 5 6 6 7 8 9
5	5 5 6 8 9

Notice that the horizontal leaves in the stem and leaf plot would correspond to the vertical bars in a histogram showing the same data, and the leaves have lengths that equal the numbers in the frequency table. From this type of graphical representation, it is obvious at a glance that there were more data in the 40s then in the 30s or 50s. However, the raw data values can still be determined. For example the bottom leaf shows that the five values in the 50s were 55, 55, 56, 58, and 59.

If you have a lot of data to enter into a stem and leaf plot, preparing it manually may introduce researcher error. You must still pay attention to details when using technology to make sure you don't make any mistakes, but using technology may save you time. If you use a spreadsheet program, you can enter your data into a column, highlight the column, and then sort the data in ascending order before you begin the construction of your stem and leaf plot.

Box and Whisker Plots

A box and whisker plot, also known as a box plot, is yet another way to display quantitative data. It displays minimums, maximums, outliers, ranges, interquartile ranges, medians, and lower and upper quartiles all in one graph. A box and whisker plot also separates the data into four parts or sections that each contains 25% of the data. Box and whisker plots are a good at comparing data groups like those in experimental and control groups.

To construct a box and whisker plot, begin by making a horizontal (or vertical) line. On this line, label values as you would a number line. Be sure to label the line with ranges that will contain both your smallest and largest pieces of data. Above this horizontal line, mark off small dashes at the lower quartile, median, and upper quartile. Make a box around these three dashes. Then make points at the minimum and maximum data pieces. Connect these points to the box by making lines, or whiskers. If there are any outliers, mark them as solitary points not connected to the box or whisker in any way.

Consider this research example: An experiment that is testing the distance a ball will travel on different surfaces includes multiple trials for each surface. For the concrete surface, the trials resulted in the following distances in centimeters (cm): 43, 49, 52, 53, 56, 58, 58, 58, 58, 63, 76. The first step is to determine lower quartile, median, and upper quartile, as shown in Figure 8.3. Then, once the medians have been identified, they can be used to graph the box and whisker plot (Figure 8.4).

Figure 8.3

Interquartile Range for Distances a Ball Traveled on Concrete

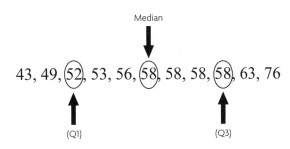

Figure 8.4

Box and Whisker Plot for Distances a Ball Traveled on Concrete

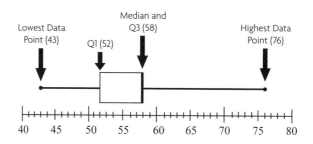

The box and whisker plot shown in Figure 8.4 is a horizontal version of a box and whisker plot, but it could have also been created vertically. The benefit of this type of graphical representation is that it concisely displays a lot of data. It shows the range of distances the ball traveled, with the shortest distance being 43 cm and the longest being 76 cm. The plot also shows that the median—which also happens to be the same number as Q3—is 58. That indicates that 25% of the numbers are this value. Another 25% of the data numbers fall between 52 and 58. The lines, or whiskers, indicate that the rest of the data numbers fall between 43–52 and 58–76.

For the same experiment, Table 8.2 shows the results of the trials completed on additional surfaces and can be used to create a box and whisker plot, as shown in Figure 8.5.

Table 8.2

Distances the Ball Traveled on Different Surfaces

Surface	Distance Ball Traveled (cm)
concrete	43, 49, 52, 53, 56, 58, 58, 58, 58, 63, 76
60-grit sandpaper	34, 36, 38, 39, 41, 45, 50, 52, 52, 52, 54
laminate	48, 51, 51, 52, 53, 55, 56, 57, 59, 60, 63
marble	24, 55, 60, 75, 79, 79, 80, 82, 82, 85, 90

Figure 8.5

Box and Whisker Plot—All Surfaces

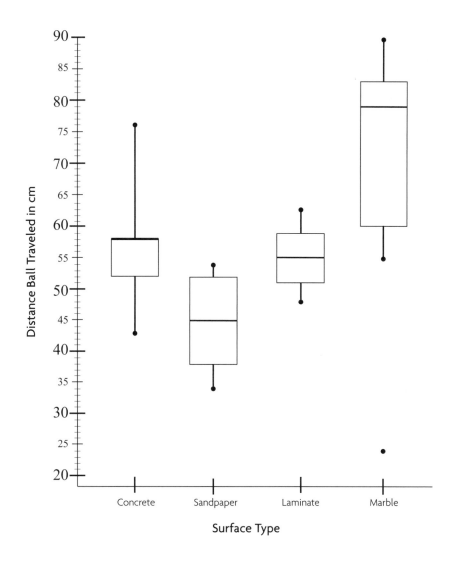

Box and whisker plots display data so that the groups can easily be compared. The axis in Figure 8.5 is oriented vertically, but it could also be constructed horizontally. The same steps are used to plot several sets of data together on a single graph that were used for plotting a single set. Notice that the shortest distance the ball traveled on the marble surface, 24 cm, although plotted, is not connected to the whisker. That is because, mathematically, it was determined to be an outlier. See the section "Dealing With Outlier Data" in Chapter 7 on pages 102-103.

Line Graphs

A line graph displays data points connected by line segments. A line graph is often used to show change in time. The Cartesian coordinate system allows you to plot points uniquely in a plane using a pair of numerical coordinates. This system displays the distances from the point to two perpendicular lines (*y*-axis and *x*-axis). You may remember this concept from algebra and geometry classes. Consider using line graphs if the measurements of an entity's change over the experiment are significant or there are drastic differences between the groups. Time is plotted on the horizontal axis or *x* axis, with your dependent variable measurements on the vertical axis or *y* axis. By using different types or colors or patterns of lines, you can compare entities of several groups at one time.

For example, Figure 8.6 shows the effect of varying amounts of wind on the length of plant stems. The time, or days of the experiment, appears on the horizontal axis, while the dependant variable of stem length appears on the vertical axis. All the experimental groups, as well as the control, can be placed on the same graph.

Figure 8.6

How Plant Stem Length Is Affected by Various Amounts of Time in the Wind

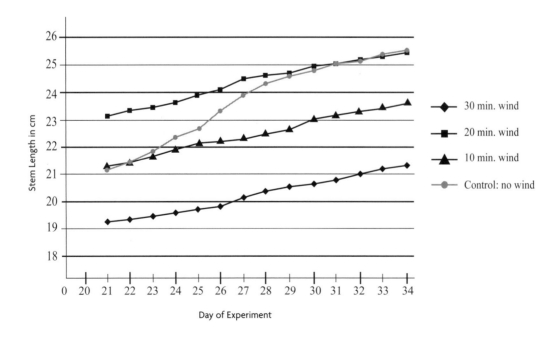

Use line graphs cautiously because they can be misleading. For example, the control group in Figure 8.6, which had no exposure to wind, grew the longest stems during the experiment represented by the steep line. However, someone might look at the graph and, noting that the 20 minutes line is the highest on the graph, might interpret this group as having grown the longest stems. Therefore, while line graphs do help show trends over time, as well as plot raw data, they are not always the best choice for graphical representation. In some cases, tables that indicate total change, or percent change, more accurately represent the data. Another misconception can arise if you make an assumption regarding the data between the data points collected and plotted in your graph. The only real data are the dotted points. It is possible that by connecting the points, you are misleading your readers. You are making an assumption—and, therefore, a prediction—about the points between the actual plotted data points. You can address this assumption when you display the data in the Results section of the final paper or poster.

Scatter Plots

Scatter plots, also known as scattergraphs, are statistical charts that plot the values of both the independent variable and dependent variable as a single point (Triola 2001). A scatter plot graphically represents a pair of coordinates (x, y) on the Cartesian plane. Scatter plots are visually similar to line graphs and help display whether or not there is a correlation between the independent and dependent variable. The independent variable is customarily plotted on the horizontal axis and shows what was measured; the dependent variable is plotted on the vertical axis. If the pattern of dots slopes from lower left to upper right, it suggests a positive correlation, and if the dots slope from upper left to lower right, it suggests a negative correlation. Patterns of dots can also indicate nonlinear correlations. The scatter plot example in Figure 8.7 is showing a correlation between the number of hours a student studies and his or her exam scores.

Figure 8.7

Positive Correlation Between Exam Scores and Number of Hours Studied

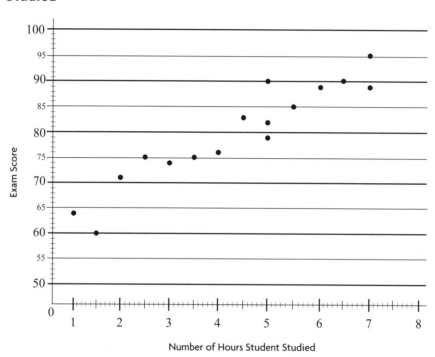

Number of Hours Student Studied

The most noticeable difference between scatter plots and line graphs is that in scatter plots the dots are not connected with a line. Notice in Figure 8.7 that for five hours, more than one dot was plotted. This indicates that three students studied five hours and each of their scores was plotted. This is an advantage of using a scatter plot over a line graph—research that has several data points for a single interval can each be plotted. Also notice that the dots are generally sloping to show a positive correlation. The closer the dots are to an imaginary line, the stronger the correlation: the more spread out the dots, the weaker the correlation.

Line graphs and scatter plots can be constructed using any graphing calculator or in spreadsheet programs using the function called XY (Scatter).

Tables

A table organizes numerical and/or descriptive data in rows and columns that aid in the understanding of a large amount of data (Cothron, Giese, and Rezba 2006). Tables are often overlooked when students begin organizing data for visual display. While histograms, box and whisker plots, line graphs, and scatter plots are fun to construct, do not underestimate the ability of a table to clearly display data. It is good practice to provide raw data in the presentation of your results (Gordon 2007). This can be done in graphs, but tables are another good way to organize this information. Table 8.3 shows a summary of the number of egg masses that were counted at various places in a pond as well as the number of confirmed fertilized eggs for each of the pond locations. Having the raw data organized in a table is helpful to readers when you begin describing the data in the Results section of your paper, poster, or presentation.

Table 8.3

Summary of Leopard Frog Egg Mass Pond Observations

Pond Identification Number	Number of Confirmed Egg Masses	Confirmation of Fertilized Eggs
38-VP-2	8	Yes
40-VP-2	31	Yes
19-VP-5	2	No
23-VP-5	0	N/A

Once raw data have been calculated via descriptive and inferential statistics, tables are helpful to compare the groups. Tables constructed with data from each of the groups help to highlight which groups had the most change and how groups compare to one another. For example, in Table 8.4 it is clear that as the level of hormone increased, so did the root growth.

Table 8.4

Total Root Growth and Rate of Growth of Plants Exposed to Different Levels of the Hormone Cytokinin

Plant	Independent Variable	Average of Total Change in Root Length Growth (in mm)	Total Time (in days)	Average Rate of Growth (in mm/day)
Control Group Plants #1–4	No hormone	8	20	0.4
Experimental Group 1 Plants #5–8	2 ml of cytokinin	14	20	0.7
Experimental Group 2 Plants #9–12	4 ml of cytokinin	22	20	1.1

Notice in Table 8.4 that groups with names like "Experimental Group 1" are also identified by how the independent variable was manipulated.

You may feel as if you keep repeating the same information in the figures, tables, and narrative. This is good! The repetition is necessary. You don't want your reader to have to flip back to an earlier page to see what each group number represents.

Tables can be made in word-processing programs or in spreadsheet programs. The data in tables and graphs must be organized in such a way that the tables and graphs can "stand on their own"—that is, no additional narrative explanation is needed. You can accomplish this by labeling rows and columns with complete descriptions. However, please note that in your research paper, you *must* mention all graphical representation of data in the narrative.

Graphical Representations for Qualitative Data

In addition to your quantitative data, you may also have collected qualitative data. Qualitative data can sometimes be graphically represented in bar graphs, pie graphs, or tables. However, it's likely that much of your qualitative data will be described in the narrative of your paper. The key is to organize the data so you can analyze the correlation between the independent and dependent variable, taking into account the qualitative variables.

Bar Graphs

Bar graphs show the distribution of qualitative data using bars that indicate frequency of a specific category. Bar graphs look similar to histograms but are best used for qualitative independent variables. For example, if you designed a categorical system to record the odor of a chemical reaction, each of the experimental groups and the control would be bars on the horizontal axis, and the vertical axis would be an odor rating of descriptions, such as *very strong, strong, noticeable, weak,* or *very weak.*

The data in Figure 8.8 is from an experiment where 50 drops of the blue reagent, Vitamin C Indicator, or dichlorophenolindophenol, were placed into test tubes. Drops of various fruit juice were added one at a time, until the blue solution turned clear. Higher numbers of fruit juice drops indicate lower levels of vitamin C. This data belongs in a bar graph, not a histogram, because the order in which the fruit juices are listed on the horizontal axis does not matter. This is because the types of fruit juice are independent of one another and are not on an incremental scale, which is how histograms are constructed. Notice, as well, that the bars in the graph do not touch.

Figure 8.8

Number of Drops of Fruit Juice It Takes to Turn Vitamin C Indicator Clear

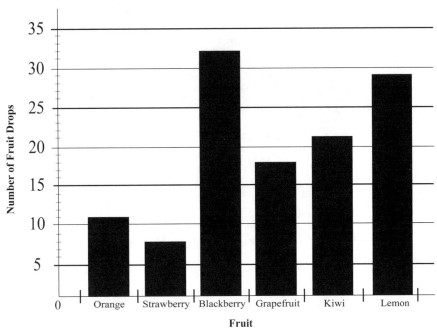

Pie Chart (or Circle Graphs)

A pie chart is a circular chart that represents data by dividing the whole circle into sectors by sizes that are proportional to the quantity the data represent. Pie charts are best for showing how data pieces fit together as a whole but are less effective for comparing data to one another. A full circle represents a whole, or 100%, and the sectors, or slices within the circle, represent the parts that make it up. For example, Figure 8.9 shows the average percent of the varying colors of M&Ms within a single bag. We understand that the circle represents a full bag, and each slice represents the percentage of a particular color in the bag. Although a pie graph is visually appealing and easy to construct in spreadsheet programs, in the case of Figure 8.9 a data table might better compare this particular set of data because each of the pie sizes is similar in size.

Figure 8.9

Average Percentage of Different Colors of M&Ms in a Single Bag

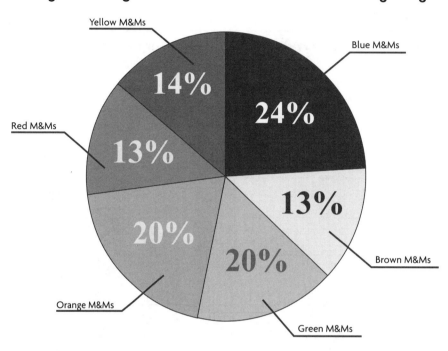

Tables and Narrative

Tables work well for qualitative as well as quantitative data. In your laboratory notebook, review your written record of procedures as well as any data tables about which you wrote a narrative. The descriptive observations you wrote on the entities themselves throughout the experiment may highlight something not captured by your quantitative data. Carefully review these narratives for patterns and trends as well as outlying and strange occurrences. This is also the time to review the photographs you took throughout the experiment. Do you notice anything different from what you recorded as quantitative data? While typing out descriptions of each entity throughout the experiment is not recommended, a table describing a condensed version of these observations would be helpful if there are noticeable trends or strange occurrences. Table 8.5 shows what a sample of what a qualitative table might look like within a research paper. Even if you decide not to insert a table such as this in your research paper, the observations you made should be discussed when you write your paper to extract additional meaning from the quantitative data.

Table 8.5

Summary of Qualitative Data and Research Journal Entries

Experiment Day	Qualitative Data and Research Journal Entries
5	After five days, there was still no mold growth on any of the bread slices. Therefore, the experimental setup was altered. To increase humidity, each bread slice was placed inside a gallon-size resealable plastic bag with a paper towel saturated with 75 ml of water.
10	Although no mold was growing on the bread slices themselves, there was a red splotch on one of the paper towels. Viewing the splotch using a magnifying glass and microscope, I found fibrous extensions of the red mass, which helped to determine that the red mass was biological, not chemical.

Making Final Graphical Representations

Using Graphical Representations to Explain the Data

The purpose of organizing your data into graphical representations is to provide a new perspective of the data. A visual representation will help you better determine what differences, if any, exist between your experimental and control groups. As you look for trends, patterns, and differences highlighted by your graphs, ask yourself questions about what you see. (See sample questions on pp. 140–142, in the "Data Interpretation" section.)

Like descriptive statistics, however, the differences highlighted by your graphs may not necessarily be the result of the changes you made to the independent variable. It is possible that these changes are due to experimental error or to chance. Therefore, inferential statistics should be calculated on the differences you notice in your graphical representations. See Chapter 9 to learn more about inferential statistics.

Reviewing the Records of Procedures and Correspondence

Another important part of organizing data is to reread your laboratory notebook sections titled "Record of Procedures" and "Record of Correspondence." This review may highlight how procedures or events affected the outcome of your results. Start by rereading your record of procedures with a highlighter and mark anything out of the ordinary. "Out of the ordinary" might be a weekend you were gone and your sister forgot to check the experiment or when someone dropped group #2 and the experimental setup had to be reconstructed. Take time to compare the dates of these unusual events with the actual quantitative data. Are there correlations between the changes in the data and the time of the event, or shortly after? If not, within the Analysis and

Conclusions section of your paper or poster, you will be able to say that these occurrences did not appear to affect the results. However, if there was a correlation, you will need to explain how and why those occurrences may have influenced the results. In addition, determine if the Record of Procedures events explains any quantitative outlier data. If so, you will need to defend your decision to keep or remove the data from the analysis. Refer to the section titled "Dealing With Outlier Data" in Chapter 7 for more information. Record your connections and inferences in the "Graphs, Statistical Analysis, and Conclusions" section of your laboratory notebook.

Consider using tables to organize any of the entries that may be important. Although you should not create a table that includes the entire Records of Procedure and Correspondence, you may want to construct a table that shows only the modifications to the proposed methods and/or occurrences that may have affected the data.

Preparing Graphical Representations for the Paper and Poster

When preparing graphical representations to include in your results and analysis and conclusions sections of your paper and/or poster, there are a few important details to keep in mind. Labels within figures and tables are extremely important. Use experiment-specific labels when describing axes and other important features within the figure and columns and rows within tables. For example, provide a description of the independent variable treatment. In Figure 8.6, the legend clearly defines each line by how the independent variable was manipulated: 30 min. of wind, 20 min. of wind, and so forth. The legend should *not* be labeled generically as "Experimental Group 1." If it were, the graph cannot be interpreted without having to look back into the narrative to see what treatments each of the groups received. Figure and table titles are also important. Each must be specific for the data it portrays. Labels must clearly describe what is contained in the graphical representation, and the horizontal and vertical axes and table columns and rows should be labeled clearly with descriptive words and units. See Chapter 11 for additional detail on how to format entire figures and tables for a paper or poster.

Chapter Applications

In your laboratory notebook, you should construct, and comment on, many different graphical representations of your data. Remember to use informative labels and detailed titles for each table or graph. Most important, make comments near the visual about your observations and your analyses of what the visual may indicate about your data, about the experiment as a whole, and specifically about the relationship between the two variables.

References

Cothron, J. H., R. N. Giese, and R. J. Rezba. 2006. *Science experiments and projects for students: Student version of students and research.* Dubuque, IA: Kendall/Hunt.

Gordon, J. C. 2007. *Planning research: A concise guide for the environmental and natural resource sciences.* New Haven: Yale University Press.

McDonald, J. H. 2009. *Handbook of biological statistics.* Baltimore: Sparky House Publishing.

Shaughnessy, J. M., B. Chance, and H. Kranendonk. 2009. *Focus in high school mathematics: Reasoning and sense making in statistics and probability.* Reston, VA: National Council of Teachers of Mathematics.

Triola, M. F. 2001. *Elementary statistics.* New York: Addison-Wesley.

Valiela, I. 2001. *Doing science: Design, analysis, and communication of scientific research.* New York: Oxford University Press.

—9—

Inferential Statistics and Data Interpretation

Introduction

Once you have completed various descriptive statistics and organized your raw data into graphical representations, as described in Chapters 7 and 8, your teacher may want you to dive further into inferential statistical analysis of that data. This chapter is organized into two parts. In the first part, I briefly introduce inferential statistical tests that are commonly used in STEM research. The description of these tests is meant only to provide an overview and to help you determine which tests might apply to your research. You will need to do additional research and decide which tests to conduct, how to perform the test, and how to interpret the results. A good resource to use is Handbook of Biological Statistics (McDonald 2009), which you can download for free (*www.lulu.com/product/5507346*). Although written for biology students, the book is easy to understand and apply to any STEM field.

In the second part of this chapter, I provide questions you should be asking yourself about the results of your experiment to interpret your data. These questions will help you as you develop conclusions about your data and determine whether or not the data support your hypothesis.

Learning Objectives

During the course of the chapter you should

1. familiarize yourself with the various inferential statistics,

2. determine which inferential statistics apply to your research data, and

3. determine whether or not differences in your data are statistically significant.

Key Terms

ANOVA (analysis of variance): An inferential statistic used to determine whether the value of a single variable differs significantly among three or more levels of a factor.

Chi-square: An inferential statistic used with qualitative data to determine if differences between frequency distributions are statistically significant.

Correlation: An inferential statistical test used to determine whether there is a statistically significant connection, or a relationship, between two variables.

Hypothesis testing: The method for testing claims made about populations; also called *test of significance.*

Inferential statistics: Mathematical calculations performed to determine whether the differences between groups are due to chance or are a result of the treatment.

Statistically significant: When differences between groups are mathematically determined to be due to the change of the independent variable rather than to luck or chance.

_t_-test: An inferential statistical test used to determine whether there is a real (statistically significant) difference between the means of two samples.

Introduction to Inferential Statistical Tests

Inferential statistics are mathematical calculations performed to determine whether the differences between groups are due to chance or are a result of the treatment (Cothron, Giese, and Rezba 2006). For example, you may have calculated means for each of your experimental and control groups, and you have noticed differences, but now you need to know if those differences are significant. The term *statistical significance* is used when the mathematical differences between groups are more likely due to the change of the independent variable than to luck or chance. Raw data alone do not indicate whether the differences between your groups are due to the treatment that you introduced in your experiment (Statistics 2011). Inferential statistics are used to mathematically determine the significance of the raw data.

You will be deciding which statistics might be useful in your experiment and then researching that method in more detail. Use statistical software (see the list on p. 98 in Chapter 7) or online statistical calculators to help you perform the statistical calculations.

Hypothesis Testing

Hypothesis testing is the "method for testing claims made about populations; also called test of significance" (Triola 2001, p. 809). Hypothesis testing consists of testing an alternative hypothesis against a null hypothesis. A null hypothesis proposes that there is no statistical significance in a set of observations. It makes the assumption that any difference in the data is due to chance, not to the treatment you introduced. Essentially, the null hypothesis is the statement you are trying to nullify, or reject, in your research. What you *really think* is the cause of the phenomenon is stated in an alternative hypothesis. It is the statistical hypothesis statement that predicts a difference between two variables and is what the experiment was set up to establish. Therefore, the hypothesis that you wrote for your proposal is actually called the *alternative hypothesis*. The alternative hypothesis is only accepted if the null hypothesis is rejected. The types of statistical tests used in hypothesis testing may include those mentioned here or others, and they depend on many things, such as sample size and type of data.

In other words, you perform inferential statistics to test the significance of your data to either support or reject the null hypothesis. If the inferential statistics determine there are significant differences between the groups you are testing, then you have evidence that the null hypothesis is not correct. This is the statistical way to show that the groups are *unusually* different and that it is unlikely that the differences are merely a matter of chance (Index 2011). You therefore would accept your alternative hypothesis.

> ## Free Online Statistical Calculators
>
> - GraphPad Software: *http://graphpad.com/quickcalcs*
>
> - Laerd Statistics: *http://statistics.laerd.com*
>
> - EasyCalculation.com: *http://easycalculation.com/statistics/statistics.php*
>
> - Stat Trek: *http://stattrek.com/Tables/StatTables.aspx*
>
> - Index of online stats calculators: *www.physics.csbsju.edu/stats/Index.html*
>
> - Interactive Statistic Calculators: *www.people.ku.edu/~preacher/calc.htm*

t-tests

A *t*-test, also known as the student *t*-test, is a statistical test used to determine whether there is a statistically significant difference between the means of two samples. In your experiment, *t*-tests might be used to calculate whether or not differences seen between the control and each experimental group are a factor of your treatment or simply a result of chance (Gonzalez-Espada 2007). For example, if your experimental design included taking height measurements, you would enter the measurements for the control group and compare them to each experimental group. You would calculate the test several times as shown below.

1st *t*-test	Control	Experimental group 1
2nd *t*-test	Control	Experimental group 2

You would conduct a *t*-test of the control group versus each experimental group in your experiment. Each *t*-test would allow you to determine whether there was any statistically significant difference between the control and that particular experimental group. If so, you could reject the null hypothesis and accept the alternative hypothesis. Using t-tests to compare control to experimental groups would allow you to make statements such as, "The mean heights from experimental group 1 were significantly higher than the control group." However, you would not be able to make any statistical comparisons between the experimental groups. For example, you could *not* state, "The mean heights from experimental group 1 are significantly different from the means of experimental group 2." To compare among experimental groups, you need additional statistical methods, such as adjusting for multiple comparisons or using an ANOVA if you have three or more treatments.

To use the t-test, each group should have at least 10 values. For example, values could be the total height that each of the 10 plant stems measured at the end of the experiment. Or the values might be the measurements that came from 10 individual trials conducted with the same conditions. Another type of t-test, called the paired t-test, is used when the two samples being compared are not randomly selected but are related, as when the same sample is measured before and after a treatment has been applied.

When you search on the internet for more information on the types of t-tests and how to calculate and interpret t-test results, use terms such as *student t-test, unpaired t-test, independent t-test*, or *paired t-test*. There are plenty of online statistic calculators that can be used for this test, such as *http://graphpad.com/quickcalcs/ttest1.cfm*.

Analysis of Variance (ANOVA)

The analysis of variance, or the ANOVA, is a statistical test used to determine whether the value of a single dependent variable differs significantly among three or more levels of an independent variable (Creech 2011). There are several types of ANOVA tests, including the one-way ANOVA, two-way ANOVA, and nested ANOVA (McDonald 2009). In context of STEM research, there are several research designs that might use an ANOVA. The one-way ANOVA works well for a single factor with three or more levels and multiple observations at each level (Experiment Resources 2011). An experiment that includes a research design that changed the quantity or quality of the independent variable can be analyzed using an ANOVA. The one-way ANOVA is an extension of the independent two-sample t-test. For example, the ANOVA test could be used to compare mean distances that a four-kilogram projectile traveled for three levels of tension on a catapult. In this example, the various levels of the independent variable are the catapult tension settings and the single dependent variable is the measurement of how far the projectile traveled. With an ANOVA test, if the results show a significant difference between the groups, this indicates only that at least two of the groups are different from each other. To determine which groups are different from which, post-hoc *t*-tests are performed.

ANOVA calculations can be done in Excel. There are numerous online tutorials that can help you perform the test in Excel, such as *www.stattutorials.com/ EXCEL/EXCEL_ANOVA.html*. There are also online calculators that can perform the ANOVA statistical test, such as *www.physics.csbsju.edu/stats/anova.html*.

Chi-Square

Chi-square is a test for "qualitative data used to determine if differences between frequency distributions are statistically significant" (Cothron, Gieze, and Rezba 2006, p. 255). If you collected *categorical data* for your experiment, you might want to use this test. (Some examples of categorical data would be viable/nonviable; male/female: simple smooth-edge leaf/simple toothed leaf/ lobed simple leaf/compound leaf/evergreen leaf; or blond hair/brown hair/ red hair/black hair.) The chi-square test will determine whether the differences in your data are statistically different from what would normally be expected.

There are two types of chi-square tests: the chi-square test of goodness-of-fit and the chi-square test of independence (McDonald 2009; Preacher 2001). The goodness-of-fit test is used to compare frequencies of your dependent variable data to known outcomes (as in genetics or when theoretical outcomes are known). For example, if your data include the number of males and females in a normal population, you would expect the ratio to be 1:1; this test will determine if the differences in your data are statistically different from the expected 50% female and 50% male ratio.

In the chi-square test of independence, you are comparing frequencies of your dependent variable data for each of your independent variables or experimental groups. These frequencies are often put into contingency tables. For example, in the river otter example used in Chapter 2, the categories of active/nonactive behavior (dependent variable) are compared for the three temperature groups (independent variable). Therefore, the contingency table for this experiment would be a 3 × 2 table.

	Active River Otter Behavior	Nonactive River Otter Behavior
Below-average temperature		
Average (expected) temperature		
Above-average temperature		

You would organize the frequency data (total number of tallied behavior) into the 3 × 2 cells, which could then be entered into an online calculator, such as *http://people.ku.edu/~preacher/chisq/chisq.htm*.

Correlation

Correlation is a statistical technique used to determine whether there is a statistically significant relationship between two variables (Triola 2001). The results of this test will help you determine whether the relationship is significant, the strength of the relationship (weak or strong), and the direction of the relationship. Before calculating a correlation test, it is best to explore the data visually using a scatter plot.

Constructing a scatter plot will help you determine whether or not a correlation exists between two variables. The visual provided by a scatterplot allows you to see whether variables move in the same or opposite directions when they change. If the variables change in the same direction, the correlation is called a *positive correlation* and if they change in opposite directions, the correlation is called a *negative correlation*. If your scatter plot looks similar to the positive or negative correlation examples in Figure 9.1, then you might calculate a correlation statistic to determine how statistically strong the correlation is.

When you search on the internet for more information on how to calculate and interpret correlation statistics, use the terms *linear correlation, Pearson product moment correlation coefficient, linear correlation coefficient,* and *linear regression*. For an online calculator, I suggest using the Pearson Correlation Coefficient Calculator at *http://statistics.laerd.com*. Note that Excel both makes scatter plots and performs the correlation test. Search for online tutorials that provide step-by-step instructions on how to enter values and interpret the results, such as *www.internet4classrooms.com/excel_scatter.htm*.

As you organize both your descriptive and inferential statistics into tables and graphs, you should be thinking about the big picture of your research. Data interpretation is the next step you will take.

Figure 9.1

Explanation of Data in Scatter plots

I. Positive Correlations Between *x* and *y*

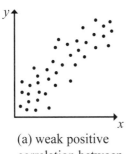

(a) weak positive correlation between *x* and *y*

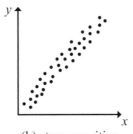

(b) strong positive correlation between *x* and *y*

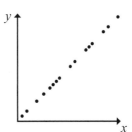

(c) perfect positive correlation between *x* and *y*

II. Negative Correlations Between *x* and *y*

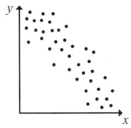

(d) weak negative correlation between *x* and *y*

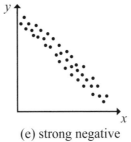

(e) strong negative correlation between *x* and *y*

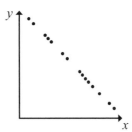

(f) perfect negative correlation between *x* and *y*

III. No Correlation

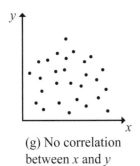

(g) No correlation between *x* and *y*

IV. No Linear Correlation

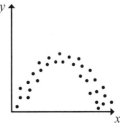

(h) No linear correlation between *x* and *y*

Data Interpretation

As you organize your raw data into tables and graphs and calculate any statistics in your laboratory notebook, you should also be evaluating the data and writing down your ideas next to the tables and graphs. You will use these evaluations to write the analysis and conclusions portion of your research paper or poster. Throughout the statistical analysis process, look at the data and keep asking yourself the following three questions.

1. What is true about my data? What additional questions come from the data?

2. How do the data describe the relationship between the two variables?

3. Do the data support the hypothesis (yes, no, or partially)?

1. What is true about my data? What additional questions come from the data?

With fresh eyes, look at the data and ask yourself questions. Below is a list of general questions to get you started. But your questions should be specific to *your* data and experiment. The answers to the questions will give you issues to address in both the results section and the analysis and conclusions section of your paper. Copy these questions and write your answers to them in your laboratory notebook.

- Why did certain groups perform better than others?

- Which group had the most drastic changes? Why?

- Why did one group do so much better than the others?

- How does the control group compare to the experimental groups?

- What *strange possible reasons* could explain the results?

- What trends or patterns are noticeable? Why did these occur?

- Are there any surprising outcomes of the experiment? Why did they occur?

- Are there some data that do not lie within the normal distribution of the majority of data? Why might this be? What could account for this?

- How might the procedure have influenced the results? (This question is huge!)

- How did any irregularities affect the results?

Use Student Handout #5, Interpreting Graphical and Statistical Data: A Peer Editing Exercise, pp. 142–144. Then, after you and your two peer editors have put together a list of questions, go find the answers. You will probably have to do more background research, and then refer to this research in the documentation of your paper/poster. There are two purposes of this peer editing exercise. The first is to get an outside perspective on your data. A fresh pair of eyes may help you see something you hadn't noticed before. The second is to provide you with practice in interpreting graphs and statistics from other research studies.

When writing the analysis and conclusions section, you will need to describe the relationship between an independent variable and the dependent variable and then explain how you know of this relationship. It all hangs on your hypothesis. What is the relationship between a and b? Describe the relationship and explain how you know the relationship exists.

2. How do the data describe the relationship between the two variables?

An important question to ask yourself is, *Did the change (independent variable) I made cause the effect that was measured (dependent variable)?* If the data support your hypothesis—but you believe it is not because of the independent variable—this distinction will need to be made.

You will answer these questions in the last few paragraphs of your analysis and conclusions section. You must either confirm or deny a relationship between the two variables and then describe the statistical data that support your final conclusion.

3. Do the data support the hypothesis (yes, no, or partially)?

Review the data you organized, graphed, calculated, and analyzed to determine whether or not your hypothesis is supported. Your answer will most likely be a simple yes or no (although it's possible that the hypothesis is partially supported). Don't write that the hypothesis was *proved* or *disproved*; instead, say that the hypothesis was *supported* or *not supported*. This distinction is important. The word *prove* is much too strong for a single study on a topic. STEM researchers prefer that a collection of research studies be carried out on a topic before saying that anything has been "proved."

It might be that all your data except one piece support the hypothesis. You may choose to make a general statement of support of your hypothesis, but in the results section in the final paper as well as the analysis and conclusions section, you will need to address exactly which data support the hypothesis and which data do not. (See Chapter 11 for details of what goes into the results section and the analysis and conclusions section.)

What if the results do NOT support your hypothesis?

Search for explanations for the results. It is likely that you will have a general sense of why you got the results you did. Either something happened during the experiment to influence the results, or maybe you now know something new about your topic and realize that the experimental design was faulty. Either way, any concrete statement about the relationship between the two variables becomes impractical.

Whatever the case, do not change your hypothesis to match the data or your data to match your hypothesis. Instead, look for possible explanations of why the results may have occurred the way they did. An important question to ask is, *Does the lack of support for my hypothesis mean that I can say that there is **not** a relationship between the independent variable and the dependent variable?* If this is a possibility, it would be valuable to mention this conclusion in the analysis and conclusions section of your paper. You could suggest another research project, with a different hypothesis, to better determine the lack of relationship between the two variables.

On the other hand, it may be that your hypothesis was not supported because other variables that were not expected influenced the results. Review your laboratory notebook. Maybe an assumption you made became an extraneous variable that you did not account for. Did any procedures you performed during the experiment adversely affect the outcome? Conduct more background research. Did you miss something that would have helped you design a better experiment or collect data more efficiently? Your conclusion and analysis section will include suggestions for future studies. You may want to suggest alternative methods or analysis for a similar research project that would retest the same hypothesis.

Remember: Having data that do not support a hypothesis is just as valuable as a supported hypothesis. An unsupported hypothesis may highlight either a possible nonrelationship between variables or an issue with the research design. *What defines a project's success is not necessarily that the hypothesis is supported but rather what it is that the researcher learns from the research process, something that he or she shares with others in the final paper or poster.*

Chapter Applications

It is important that you research various statistical tests and determine which ones to use on your data. Calculate statistics and write the results in your laboratory notebook. For each statistical test, answer the three questions on page 140 to help analyze your data.

At this point in the handbook, we detour briefly from "science talk" to "documentation"—that is, the rules you need to follow to make sure you do not plagiarize. Chapter 10 explains how to properly give credit to the resources you used in your background research and that you will be referencing in your STEM research paper, poster, and/or presentation.

References

Cothron, J. H., R. N. Giese, and R. J. Rezba. 2006. *Science experiments and projects for students: Student version of students and research*. Dubuque, IA: Kendall/Hunt.

Creech, S. 2011. Statistically significant consulting. Retrieved May 28, 2011, from *www.statisticallysignificantconsulting.com*.

Experiment Resources. 2011. Statistics tutorial. Retrieved May 21, 2011, from *www.experiment-resources.com/statistics-tutorial.html*.

Gonzalez-Espada, W. 2007. Using simple statistics to ensure science-fair success. *Science Scope* 8 (30): 48–50.

Index of online stats calculators. 2011. Retrieved May 30, 2011, from the College of Saint Benedict and Saint John's University's physics website: *ww.physics.csbsju.edu/stats/Index.html*.

McDonald, J. H. 2009. *Handbook of biological statistics*. Baltimore: Sparky House Publishing.

Preacher, K. J. 2001. Calculation for the chi-square test: An interactive calculation tool for chi-square tests of goodness of fit and independence [Computer software]. Available from *http://quantpsy.org*.

Statistics in science fair projects. 2011. Retrieved May 20, 2011, from *www.scitechfestival.com/pdf/statisticsdocument.pdf*.

Triola, M. F. 2001. *Elementary statistics*. New York: Addison-Wesley.

Interpreting Graphical and Statistical Data

A Peer Editing Exercise

(This exercise should be completed by two peer editors.)

Researcher's Name _____ Date _____

Peer Editor's Name _____ Date _____

Part 1

Directions for the Researcher:

- Consider completing this student handout online so you can upload graphs and proposals and the two peer editors can comment electronically.

- Organize the data from your research into graphical form (tables, bar graphs, line graphs), assigning each figure its own number and title (*Example:* Figure 1: Amount of Frost at Different Hours of the Day). Make a copy of a complete set of the figures for each of your peer editors.

- Complete Part 1 of this handout. Give a copy to each of the peer editors.

- Have your peers answer the questions in Part 2. Complete Part 2 yourself as well—you'll be both the researcher and the editor.

What calculations did you make with the raw data you collected to develop the data in these tables and figures (your figures will probably be graphs; graphs are a kind of figure)? Write at least one paragraph, using complete sentences.

_____

_____

_____

_____

_____

_____

Hypothesis:

_____

_____

_____

_____

Part 2
Directions for the Peer Editor:

State the Facts

Look at the graphical and statistical data: State at least 10 facts that the data show (e.g., "The experimental group that contained the most sulfur had the most change.")

1. _____

_____

2. _____

_____

3. _____

_____

4. _____

_____

5. _____

_____

6. _____

_____

7. _____

_____

8. _____

_____

9. _____

_____

10. _____

_____

Ask the Hard Questions

When you looked at the graphical data and wrote down the facts above, what questions came into your mind? Your question might be simply "Why?" regarding any of the facts above, but more helpful to the researcher would be questions such as "How might the methods have influenced the results?" or "How were the raw data organized for analysis?" Your questions and comments should be specific to the study. You might also comment on any assumptions you think the researcher made. Ask the researcher about external variables that might have influenced the results and how those might be addressed.

Was the Hypothesis Supported?

State whether or not the researcher's experiment supported a correlation between the two variables (yes, no, partially). Then explain your statement in no more than three sentences. Use the data in your explanation.

Comment on the Tables, Graphs, and Statistics

Please make suggestions for possible improvement of the graphs. You could suggest changing the wording or location of labels, the wording of titles, or the types of graphs that were chosen to depict the data. If you have any technology tips on how to make the graphs easier to read, share those. Make these suggestions on the graph sheets themselves.

— 10 —

Documentation and Research Paper Setup

Introduction

After organizing the raw data from your experiment and performing statistical analysis, you are ready to move to the next part of your research project—sharing the results of your experiment. You do this by writing a STEM research paper and/or preparing a poster and presentation. In a paper or a poster, you must properly document, or cite, the resources you used. In other words, you must document within your paper and in a list at the end of the paper the resources (e.g., journals, online material, books) you used to do the research project and paper.

Learning Objectives

By the end of the chapter, you should be able to

1. explain what documentation is and why it is important,

2. list the documentation guidelines that apply to all documentation styles, and

3. describe the connection between in-text citations and the Works Cited list.

Key Terms

Citation: A brief description of a resource placed within a piece of writing. A fuller description of the citation is found in a list at the end of the piece of writing. *Citation* and *reference* mean basically the same thing.

Documentation: The practice of giving the source (e.g., the book, article, internet) of something (e.g., a fact, opinion, or quotation) that you are using in your own writing.

Other works consulted: A list of resources you used when researching your paper but did not cite in the paper. (Not all papers will have this list.) Use the same style as for the Works Cited list.

Works cited: The list of resources cited in a piece of writing when using MLA style documentation. Other documentation styles call this list "References" or "Bibliography."

A research paper must follow the documentation style of an official style guide. In this book, you are learning to use the documentation style of the Modern Language Association (MLA). It is true that college students or college teachers and researchers who write STEM research papers do not use MLA style documentation (see Chapter 3, page 40, footnote, for my reasons for choosing this style). In college, you will most likely use the American Psychological Association (APA) style of documentation or one more specific to your field. For example, the field of physics commonly uses the American Institute of Physics (AIP) style manual, the fields of chemistry and environmental science use the American Chemical Society (ACS) style manual, and the medical field uses the American Medical Association (AMA) style manual. NSTA Press, the publisher of this book, uses the *Chicago Manual of Style*.

I am introducing you to the general principles of documentation. You should find it easy to transfer the basics of what you learn here to any other documentation style.* You will find more information about MLA style and other official documentation styles at the following sites:

- Purdue Online Writing Lab (OWL) *http://owl.english.purdue.edu/owl/resource/747/01/*

- Noodle Bib *www.noodletools.com*

- Study Guide *www.studyguide.org/MLAdocumentation.htm*

There are several important general principles of documentation. The most important is that you give credit to the ideas, information, or expressions of other people in two places within your scientific paper or poster: the narrative (the essay part) of your writing and in a reference list (called Works Cited here) at the end of your paper. All documentation styles require in-text citations. *Any sentence in your paper that comes from ideas you got during your background research*

---

* The MLA instructions provided in this chapter and elsewhere in the book are based on the seventh edition of the Modern Language Association's style handbook (*MLA handbook for writers of research papers*. 2009. New York: Modern Language Association of America).

must have an in-text citation. In MLA style, you put the author(s) name(s) and page number(s) (i.e., the pages where the idea or quote appeared) within parentheses: (Smith 124).

The in-text citation is a signal to your reader to refer to the Works Cited list so that he or she has a way to obtain the source you cited. As noted earlier, in MLA style documentation, the list of resources at the end of your paper is called a Works Cited list. Any resources that you used but did not cite in your paper can be put into an Other Works Consulted list.

Three Aspects of MLA Documentation

The box to the right gives you the five general principles of documentation. Now let's move on to the specifics of MLA style documentation.

General Principles of Documentation

(Rules That Apply No Matter What Documentation Style You Are Using)

- Place an in-text reference within the narrative in parentheses. Put it at the end of a sentence or sentences that contain ideas from other sources, such as books or journals.

- Include complete information about the source in a Works Cited list/bibliography/reference list.

- Alphabetize that list by the author's last name.

- In-text references must be listed in the Works Cited list, and the entries in the Works Cited list should be cited somewhere in the paper. Needless to say, the spelling of an author's name within the text and in the Works Cited list must match!

- Punctuation, spacing, and abbreviations of in-text citations as well as the reference list are important and specific to each documentation style.

1. Citing Sources Within the Paper (Author Last Name and Page Number)

Citing your sources within the text of your paper means inserting citations in parentheses right after the idea or quote you borrowed from another source. There are two primary ways to document or cite sources, but both must include the author's last name(s) *and* page number.

- Put the author's last name and the page number in parentheses after the sentence where you used the source. This information should all be in your notes (see Chapter 3, pp. 35–36). It is not a good idea to write the paper first, with the intention of going back and inserting the in-text citations later. Include them as part of the writing process, not later when there is an increased chance that you might not accurately cite your references (or maybe forget to cite them altogether!).

- Punctuation, spacing, and abbreviations are very important when making in-text parenthetical citations. In MLA style documentation

(Figure 10.1), there is one space between the last word in the sentence and the opening parenthesis. However, there is no space between this parenthesis and the first letter of the author's last name. Insert one space between the last letter of the author's last name and the page number. Notice that there is no abbreviation of the word *page*—only the number by itself. After you close the parenthesis directly after the page number, put the punctuation for the entire sentence *after* the closing parenthesis. (*Note:* These rules may seem incredibly picky, but if they aren't followed consistently throughout your paper, the paper will look sloppy and unprofessional.)

- Mention the author's last name inside the sentence and place only the page number(s) of the source in parentheses (see Figure 10.2). If the work you are citing has no known author, use either the complete title or a short form of the title of the book or article within the parentheses. If you are using a book title or article title instead of an author, remember that book titles should be *italicized* and titles of articles are put in "quotation marks." The first letter of what you put in parentheses (not counting *A, And,* or *The*) determines where that entity will appear in the (alphabetized) Works Cited list. If the work you are citing has more than one author or you run into other documentation issues (and you probably will), check the MLA style handbook itself if your teacher has a copy or go to *http://owl.english.purdue.edu/owl/resource/747/01.*

Figure 10.1

MLA Style Citation With Author's Name at End of Sentence

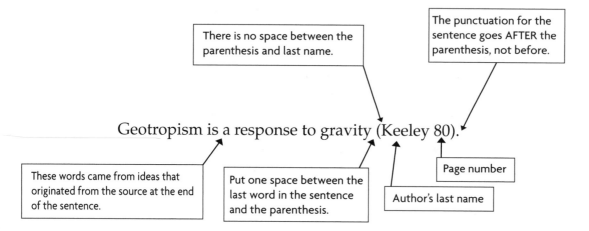

There is no space between the parenthesis and last name.

The punctuation for the sentence goes AFTER the parenthesis, not before.

Geotropism is a response to gravity (Keeley 80).

These words came from ideas that originated from the source at the end of the sentence.

Put one space between the last word in the sentence and the parenthesis.

Author's last name

Page number

Figure 10.2

MLA Style Citation with Author's Name Inside the Sentence

▼Keeley notes that only multicellular organisms can produce eggs (118).

| Author's last name is used as the subject of the sentence. | Name left out of parentheses and only page number remains. | Punctuation is still after the parenthesis. |

2. The Works Cited List

The second part of documentation is the creation of the Works Cited list. This list includes all the resources you cited in your paper. Below are the basics of how books, journals, and online resources should be formatted on the Works Cited page. Use the same formatting if you have an Other Works Consulted list.

Book With One Author

Last name, First name. *Title of Book*. Place of Publication: Publisher, Year of Publication.

Example: West, Herbert. *Forms of Energy*. New York: Rosen Publishing Group, Inc., 2009.

Book With More Than One Author

Last name, First name, First name Last name, and First name Last name. *Title of Book*. Place of Publication: Publisher, Year of Publication.

Example: Silverstein, Alvin, Virginia Silverstein, and Laura Silverstein Nunn. *Matter*. Minneapolis: Twenty-First Century Books, 2009.

Article in a Magazine or Journal

Last name, First name. "Title of article." *Title of Magazine or Journal* Volume.Issue (Year): pages

Example: Ruxton, Graeme D. "Zoology: Why are whales big?" *Nature* 469.7331 (2011): 481.

Entire Website/Editor, Author, or Compiler (If Available)

Editor, author, or compiler name [if available]. *Name of Site*. Version number. Name of institution/organization affiliated with the site [sponsor or publisher], date of resource creation [if available]. Medium of publication [i.e., print or web]. Date of access. [URL if required by your teacher]

Example: Jones, John. *How Stuff Works.* Discovery, 1998. Web. 10 Oct. 2011.

A Page on a Website/No Editor, Author, or Compiler Available

"Title of Article/Webpage." *Name of Site.* Version number. Name of institution/organization affiliated with the site [sponsor or publisher], date of resource creation [if available]. Medium of publication [i.e., print or web]. Date of access. [URL if required by your teacher]

Example: "How Do Flies Breathe?" *How Stuff Works.* Discovery, 22 Jan. 2001. Web. 31 March 2011.

When typing up your Works Cited list, remember these basics:

- The Works Cited list should have the same 1 in. margins as the rest of the paper and be numbered consecutively with header formatting to match the rest of the paper.

- Note the *punctuation* in the MLA Works Cited because its placement is significant. Every period, comma, colon, and space is important.

- Longer works, such as books and magazines, are *italicized.* Shorter works, such as articles, are put in "quotation marks."

- The words Works Cited are centered on the first line of the page. Don't use any fancy fonts, quotation marks, italics, bolding, or underlining.

- Alphabetize the Works Cited by author's last name. If there is no author, alphabetize by the first major word in the title of the work. Never alphabetize by articles like *a, and,* or *the.*

- The entire Works Cited is double-spaced, with what is called a *hanging indent.* This means that the first line of the entry is aligned with the left margin, but if the entry is more than one line, then the following lines are indented five spaces or about ¼ in. (the line that is indented is the "hanging indent"). Making the indent allows the author's name to be seen easily.

3. Matching Paper Citations With the Works Cited

Your final step regarding documentation is to make sure that the in-text citations that are in parentheses match the references in the Works Cited list. Your reader should be able to quickly flip from the mention of resource in the text to the Works Cited list to find complete information about the resource.

Amount of Documentation and Use of Quotations

You may wonder, as students often do, just how much documentation to use in a research paper. Most teachers would prefer that you "over document," rather than "under document." When in doubt, by all means insert another in-text citation. Ideally, you should have at least two different resources backing up your points in every paragraph. (I emphasize different because if you depend too heavily on one resource per topic, your reader might question the authenticity of what you're saying and think, *Only one person seems to believe that. Maybe it's just that person's opinion and isn't really factual.*)

A quotation, or direct quotation, is the use of the *exact words* from a resource. Use quotations sparingly. When you use them, don't let them stand alone without some explanation or discussion by you. Show the reader that you used the quotation to support an idea you are trying to get across. (See pp. 48–49 for more information on how to use quotations in your paper.)

Research Paper Setup

The First Page of the Paper

In MLA style, there is no separate title page. The actual (written) paper begins on the first page after you follow these four steps:

Step 1. On the first page, put the following information in the upper left-hand corner, double-spaced.

- Your full name

- Teacher's name

- Course name

- Due date

Step 2. On the first page, double-space and center the title of your paper. Don't put it in quotation marks or bold face and don't use all capital letters.

> **Example:** The Effects of Glucose on Cell Volume

Step 3. Plan to put the "header" or "running head" in the upper right-hand corner of every page. It should your last name and page number, with a space in between: Harland 12

Step 4. Plan to double space the entire paper. Be sure to never hit the return twice between section headings or between references listed in your Works Cited. The margins should be set to 1 in. on all sides. (Be careful: Often word processing programs have a 1.25 in. default setting). The font should be Times New Roman or Arial in 12 point size, with no other fonts used anywhere in the paper.

NOW, you begin the text of your paper, double-spaced under the title.

Subdivisions Within the Paper

Subheadings should be *centered*. Use the same font and type size as for the rest of the paper (and, again, no bold, underling, italics, or all capitalization). Each heading should be followed by at least two paragraphs within them.

Formatting the Works Cited

The entire page is double-spaced, with the first line of each entry being left-justified and any following lines allowing a hanging indent of five spaces, or ¼-in. The entries are not numbered. They are listed in alphabetical order by author's last name or first main word in title. Follow the same format for you Other Words Consulted list, if you have one.

Tables and Figures

According to MLA style documentation, all graphics inserted into papers are divided into two categories; tables or figures. Both tables and figures must remain within the 1 in. margin of the paper. (*Note:* All tables and figures must be referred to in the text. Also, the tables and figures should be numbered separately—Table 1, Table 2, etc.; Figure 1, Figure 2, etc. This means that it would be possible, for example, to have both a Table 4 and a Figure 4 on the same page.)

Tables

Your STEM research paper may include tables that display experimental data or statistical results. You will probably construct tables either within the word-processing program you are using or by copying and pasting from a spreadsheet program. It is important to place the table as close to the text to which it relates. Each table is titled: Use the word *Table* and its assigned number on the first line and then a descriptive title on the second line. The lines are aligned at the left and double-spaced. The numbered table head and its title are placed *above* the table in the paper (not under the table as is the case for figures) and left justified. For example:

Table 6

Comparison of Some Physical Properties of Carbon Allotropes

Figures

In MLA style documentation, all visuals or graphics that are not classified as a table are considered figures. For your STEM research paper, figures might include any graphs of your experimental results or photographs you are choosing to include. Each figure is labeled "Figure," its assigned number, and a descriptive title. For example:

Figure 4. *E. coli* Growth Within Varying Concentrations of Streptomycin.

The figure label goes under the figure, unlike the table label, which goes above the table.

Chapter Questions

1. What is documentation and why is it important?

2. What basic documentation guidelines apply to all documentation styles?

3. Describe the connection between in-text citations and the Works Cited list.

Chapter Applications

This chapter provided enough background on how to use the MLA style documentation to get you started in correctly formatting your resources during the background research phase. While you take notes, use the MLA Works Cited section in this chapter (p. 151) to format resources on bibliography cards, notebook pages, or online. Once you are ready to begin writing and formatting your paper or poster, use this chapter in conjunction with the three MLA style documentation resources listed on page 148. Remember to insert in-text citations as you write, not after, and to match the citations to the Works Cited list.

Before you can actually begin writing, however, you will need to organize your STEM research paper into the standard scientific journal form. The next chapter will show you how.

-11-

Writing the STEM Research Paper

Introduction

You have completed your STEM research experiment and performed statistical analysis. It is now time to organize your information so that you can write the paper or design a poster. To begin, you may want to create an outline or a graphic organizer (e.g., a chart or concept map) to help you determine what you will put in each section. The STEM research paper or poster you are going to write will be organized into five sections: Introduction, Materials and Methods, Results, Analysis and Conclusions, and Works Cited. You might also have one or more appendixes. Before you begin writing, please review the plagiarism precautions found in Chapter 3 (pp. 47–48).

Learning Objective

By the end of the chapter, you will be able to list and describe the five sections of the STEM Research Paper.

Key Terms

Abstract: A single paragraph that summarizes a research project. Its purpose is to help readers quickly and easily decide whether or not they want to read the paper. Abstracts include what was studied, how it was studied, the results, and a brief analysis of those results.

Analysis and Conclusions: The section of a scientific paper that interprets the data that were reported in the Results section.

Introduction: The section of a scientific paper that states the problem or project topic and tells why the problem or topic is being studied in an experiment. The introduction should include documented background research on the entity being studied, the independent and dependent variables of the experiments, and the various methods that will be used.

Limitations: The part of the Analysis and Conclusions section that discusses anything about the experiment that may weaken the confidence level of the results. Limitations may include extraneous variables not kept constant, poor sampling, or instrument errors.

Materials and Methods: The portion of a scientific paper that tells how the study was conducted, what equipment and techniques were used, and what procedures were followed. The description of the procedures should be detailed enough so that so that someone else could replicate the experiment exactly.

Results: The section of a scientific paper that reports on the experimental findings of the research study, including the statistical analysis of the findings. It should include both graphics and a written account.

If you performed the experiment as a group, your teacher may still expect each group member to write an individual STEM research paper. If, on the other hand, your teacher expects to receive a single paper that represents the work of the group, that paper must flow as if it were written by a single person (even if parts of it were written by different group members). In particular, take care to ensure that the wording between sections (the transitions) is smooth and that the paper reads as one cohesive piece. Don't just copy and paste each group member's portion into the same word-processing file the night before the paper is due. Instead, each group member should write his or her pieces in a single document that everyone edits, such as a Google Doc.

Some groups find that writing the entire paper together in Google Docs is more constructive than dividing the writing among members. It is up to the teacher or group to decide. In either case, *before you begin the paper*, write up a contract agreed on by all group members that clearly establishes everyone's roles.

In any case, remember to back up your electronic files. Every couple of days save the file(s) in at least two places, perhaps on a computer hard drive, disk, or flash/thumb drive. You can also back up files by uploading a copy online—using services that will store files online for free (e.g., Dropbox at *www.dropbox.com*) or uploading a new copy each week to Google Docs. Another simple backup procedure is to send e-mail attachments of the file to yourself or a friend. Your friend could keep the file in his or her inbox or save it to the hard drive. You will be spending too much time on this piece of work to rewrite it. Back up your files regularly.

11

Parts of the STEM Research Paper

Your STEM research paper will have five (or more) sections: Introduction, Materials and Methods, Results, Analysis and Conclusions, and Works Cited and maybe an abstract and appendixes. When typing your paper, you should center these headings on the page, without any special bolding, italics, or underlining. Don't use all capital letters; copy the style that is used in this paragraph.

Introduction

As noted in Chapter 5, when you were working on writing your proposal, the purpose of the Introduction is to state a problem or project topic and why you are studying it. You should explain how the experiment you designed addresses the larger problem. An Introduction is more than just a summary of your background research; it should address the questions you researched while developing your research design. In other words, the Introduction should weave together your background research on the independent and dependent variables with your data collection methods. The Introduction should show the path you took to address the problem. As a whole, it should move from general to more specific.

Write the Introduction as if it were an essay that could stand on its own. Figure 11.1 (p. 160) will help you organize this part of the paper. The inverted triangle shown in Figure 11.1 represents the first paragraph(s) of the Introduction. The opening sentences of these first paragraphs should be written in broad terms, explaining the context of the problem that your research study will address as well as the importance of the study to the scientific community. The last sentence of paragraph 1 of the Introduction should *be a version of the hypothesis*. For example, that sentence might be something like, "Therefore, this research study was conducted to look for a relationship between ____ and ____."

Once you have written the first paragraph(s) of the Introduction (i.e., the "Describe the Problem" part), you organize the body of Introduction. Each rectangle in Figure 11.1 represents a part of the Introduction. The key is to begin each paragraph with a topic sentence that explains what that paragraph will talk about. As noted in Figure 11.1, write a paragraph(s) on the entity studied. Describe the care needed for the entity and any safety or other common issues that can arise when working with this entity. Also include paragraphs on the independent and dependent variables and, in the context of your hypothesis (which you stated earlier in the last sentence of the first paragraph), explain why these variables were chosen.

Figure 11.1

Paragraph Organization for Introduction of a STEM Research Paper

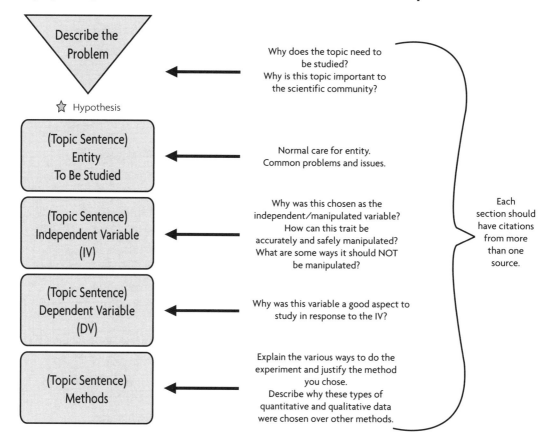

In addition, cite studies you found during your background research that were similar to your study (put the name of the author of each study in parentheses; make sure that that author also is listed in the Works Cited section). Explain how the other studies differed from what you were trying to accomplish. Finally, explain the method you used, possibly defending why it was used if there are several methods that could have been used.

Refer to the MLA documentation chapter (Chapter 10) as well as to the suggestions provided for scientific writing in the proposal chapter (Chapter 5, especially pp. 73–75). Use your notes, not the original sources, as you write. Be sure to cite more than one source per paragraph. This increases the reader's confidence that your research was thorough—that is, that you verified the information you found in one place with the work of a different researcher. Rely on your teacher's guidelines regarding whether active or passive voice is preferred and whether pronouns are appropriate in the introduction (see p. 74).

Materials and Methods

You will combine three sections from your proposal—Hypothesis, Materials, and Methods (see Chapter 5)—into one section—Materials and Methods—in the final paper. In the Materials and Methods section, you describe how you conducted the study, what equipment and techniques you used, and what procedures you followed. Use your proposal as a starting place for this paper.

If, when you were writing your proposal, your teacher had you write the Methods section in paragraph (narrative) form, you will only need to make some grammatical changes (e.g., perhaps change the future tense to past tense—for example, the statement, "I will heat the solution to 40°C" would be changed to "I heated the solution to 40°C.") and update the Methods section to communicate what actually occurred. However, if you wrote a numbered list of experimental procedures for your proposal, you will need to rewrite this section into narrative form.

The hypothesis that your teacher approved in your proposal should be integrated somewhere into this section. You could begin the section by stating what the study was designed to test, then give the exact wording of your hypothesis. (However, you might find the hypothesis fits better at the end of this section.)

A separate materials list is not needed for the final paper. Instead, just make sure that *each item you actually used* is mentioned in the Materials and Methods section.

The Materials and Methods section is also the place to insert photos of the experimental setup and photos of how data were collected. Each photo is considered to be a "figure" in the paper, so give each photo a figure number and title and mention it in your narrative, directing the reader to look at the figure—say, for example, "The pH was measured using a probe, as shown in figure 4. (See the Results section later in this chapter for more information on putting tables and figures into MLA style.)

Although you will not be including the entire Experimental Design Table (from Chapter 2) from your proposal in you final paper, you could modify it into a smaller table to compare your experimental groups with your control group (see Table 11.1). If you choose to do this, be sure to refer to it in your narrative.

Table 11.1

Group Description for Experimental Design

Control Group	Experimental Group #1	Experimental Group #2	Experimental Group #3
No potassium (K)	5 g	10 g	15 g

Results

The purpose of the Results section of the paper is to provide your reader with the experimental findings of your research study, including a statistical analysis. Do not, at this point, attempt to explain what the data mean. In the Results section, you just describe the results without making *any* judgment as to what the results may, or may not, indicate. Therefore, graphical representations of the data paired with text will make up this section. (Refer to Chapters 7, 8, and 9—on statistical analysis—for discussions of how to present the specific types of data you analyzed.)

Organizing the Results Paragraphs

Start this section with an explanation of how you prepared the data for analysis. Tell what mathematical computations and statistics you performed on the raw data to prepare the data for analysis. The purpose of this explanation is to be transparent—sharing how you prepared the data for analysis allows for both a critical review by others and a possible replication and analysis in exactly the same way.

Your laboratory notebook contains many more tables, graphs, and diagrams than you will actually put into your paper. What you write about in the Results section will be based on the graphic images you choose to include in the final paper or poster. Use strong topic sentences to signal to your reader what data you will be presenting. These paragraphs, when organized well, should make clear the trends and patterns in the data. (It is possible that you may have outliers—that is, points in a sample that are widely separated from the main cluster of points—in the data. Do not ignore these. Instead, address them plainly, without giving possible explanations for why they happened.)

When referring to data from specific days of the experiment, do not use the actual dates because they don't give your reader any indication of how that data compares to other data in the experiment. Instead, refer to the *day* of the experiment. For example you would say "on day 17 of the experiment…" not "on February 2…." Calendar dates are in your laboratory notebook for your own reference, but do not use them in the narrative. However, you can provide the beginning and ending dates of the experiment for general reference. For instance, in an experiment relating to weather conditions, knowing the particular date or time of year might be especially important in analyzing or replicating the experiment.

The Results section is written in the past tense. As for which "voice" to use, check with your teacher as to whether active or passive voice is preferred and whether personal pronouns should be used (refer to pp. 73–75). Some readers find the use of the pronoun "I" distracting—for example, "I then took the raw data for each group and calculated the mean." In the passive voice, that sentence would read, "A mean for each group was calculated." As you can see, a sentence constructed in passive voice reduces the emphasis on who did the calculations.

Preparing Tables and Figures for the Results Section

As explained in Chapter 10 on MLA style documentation, all visuals or graphics that are not classified as tables are called *figures*. Each table and figure must have an assigned number as well as a title that clearly describes what it displays. Tables are numbered consecutively, in the order they appear in the text. The same is true of figures. Tables and figures, however, are numbered independently of each other. For example, on the same page you could have both Table 3 and Figure 3.

Each table and figure number should appear in two places: in the title of the table or figure and in the text that refers to the table or figure. Also, tables and figures must appear in the order that you refer to them in your paper. For example, if Table 1 is a table comparing the control and experimental groups, you must refer to it in the text and it should appear on the page *before* Table 2 appears.

When referring to tables and figures, do not capitalize the *t* or the *f*. (You might have noticed that in this book the *t* in *table* and the *f* in *figure* ARE capitalized. That is because the publisher of this book uses the *Chicago Manual of Style*, which has different rules from the MLA style handbook.). So, in your paper, an in-text reference might read, "The dew point spiked to 89.7 on day 12 of the experiment, but leveled out between days 15–20 (see figure 1)."

When you are referring to tables and figures in the text of your paper, refer to specific aspects within the table or graph, not just to its number. See sidebar for specific examples.

Any table or figure too big to put in the text of the paper, such as a long questionnaire given to study participants, can be put in an appendix that is

---

How to Refer to Tables and Figures in a Research Paper

Poor Example: There were three mathematical patterns found. See figure 6.

Good Example: In all six sets of sheet music analyzed for this study, all but one showed a pattern between the speed at which the music was to be played and the number of notes. As figure 6 shows, the number of musical notes in measures to be played at *adagio* was consistently half of those to be played at *presto*. However, one musical piece, shown by the dotted line in figure 6, showed no such relationship between speed and number of notes.

Poor Example: All the algae groups grew over the course of the experiment. See figure 3.

Good Example: The algae group with the highest acidity levels grew an average of 3 cm² in the first week of the experiment. But as the sharp decline in figure 3 shows, the total surface area decreased, ending with only 2 cm² growth at the conclusion of the experiment.

placed after the Works Cited. These tables or figures must be referred to in the paper even though they appear as appendixes. Number them in the order they are referred to in the paper (Appendix 1, Appendix 2, etc.). Appendixes are considered a part of the paper. Also, any written permissions you had to get in order to do vertebrate or human subject research are considered appendixes. Each is given its own number (e.g., Appendix 3) and is referred to in the paper.

Tables and figures are the keys to well-done Results and Analysis and Conclusions sections. When your teacher is evaluating your paper, he or she will use the visuals in the Results section to look for patterns, strange occurrences, and large and/or small changes in the data. He or she will then check to make sure that you have discussed these occurrences in the Analysis and Conclusions section.

Analysis and Conclusions

The purpose of the Analysis and Conclusions section is to explain the data you reported in the Results section. (In some STEM journals, this section is called the Discussion.) It can be the hardest section to write because you must interpret your results and draw conclusions, processes you might not have a lot of experience with. An important component of the Analysis and Conclusions section is that you declare the hypothesis supported or not supported or partially supported (Day and Gastel 2006).

Introductory Paragraph to the Analysis and Conclusions Section

Begin this section by stating whether or not the hypothesis was supported and making general comments as to how strongly it was (or was not) supported. Then list your explanations for this finding that you will be discussing in the rest of this section. You explain the evidence to support these claims in the supporting paragraphs. Ask your teacher if you are to use first person in this section. If you are, the sentence can read, "My hypothesis was not supported because…." However, if your teacher prefers that you stay neutral in your remarks, that sentence might be worded more like, "The hypothesis was not supported because the water level of retention ponds varied more throughout the spring than all the other bodies of water in the study."

Next, list all of your possible explanations: "This may have occurred because a, b, c, d, e, or f." (This sentence, or sentences, acts like a thesis statement in an essay.) Then you address the reason you gave as "a" in its own paragraph. Next you address "b" in the next paragraph, and so on. You might prefer to write all the explanation paragraphs first, decide what order they should be presented in, and then write the thesis statement ("This may have occurred because a, b, c, d, e, or f").

Paragraphs in the Analysis and Conclusions Sections

Each paragraph should address *one* aspect of the explanation of the results. To help the reader through this section, topic sentences should clearly tell him or her what will be discussed in each paragraph.

Because all the data were presented in the Results section, you only need to restate the results that you want to comment on. It is appropriate to refer the reader back to tables and figures in the Results section. When explaining the results, don't use words such as *obviously, clearly,* or *proves.* The words *obviously* and *clearly* are insulting to your reader (who, you should assume, doesn't need to have the obviousness of something pointed out to him or her), and *proves* is too strong a word for a single study.

When appropriate, discuss any groups that had irregular results compared to the rest of the groups. Try to explain why this might have happened. Also, in your explanation of the results, be sure to address questions that were posed by your peers in the peer editing exercise you completed in Chapter 6.

All the trends and patterns you reported in the Results section must be explained. In other words, you must answer the question, "Why did that happen?" Most important, you will need to do all of this—explain your results— by citing past research that is documented according to MLA style. Here, of course, you will have to go back to your background research. All scientific facts MUST be documented, not assumed. For example:

> *During the fall, hydras reproduce sexually (Lentz 13). Even if conditions were right for sexual reproduction and two hydras—such as the two in culture 15—started to reproduce sexually, this would not show up in this experiment because a fertilized hydra egg can take three to six weeks to hatch (Lenhoff 2).*

It is also appropriate to say that your methods might have influenced the results. However, you need to do more than just suggest a possible influence; you must explain *how* the methods may have influenced the results and what could have been done to prevent that influence. For example:

> *Despite my efforts to control the amount and intensity of light exposure during the experiment, I was not as careful about monitoring the light for each of the specimens while collecting data each afternoon. Some specimens were out of their controlled lighting setup for longer periods of time than other specimens. The additional variable of exposure outside the light setup may have influenced the results. This is particularly true for the experimental group that was to be exposed to no light. In future experiments, time to collect data should be equivalent and monitored.*

Other Topics to Be Included in the Analysis and Conclusions

Limitations

After you have discussed all explanations of the results, you need a paragraph on the limitations of the study. *Limitations* are aspects of the research that may weaken the confidence level of the results. For example, maybe there were variables you were not able to keep constant, and therefore, extraneous variables may have influenced the results. Or there were problems that occurred during the study that limit your ability to apply the results to a more general conclusion. Or the number of trials or number of data collection days may not have been sufficient to apply the results beyond this study. Address limitations within the body paragraphs of this section as they apply to specific aspects of the research or you address them in a paragraph of its own at the end of the Analysis and Conclusions section.

The Analysis and Conclusions should connect back to your Introduction. You chose to do this study to address a general question you had. Now you need to connect whether or not your research study provided any answers to that original question. You may also want to discuss the possible applications and extensions of your research study. Describe possible research studies that could be completed in the future. These suggestions might be slight modifications of your own study or extensions that could be completed to answer new questions brought up by your research study.

Apply the results of your experiment globally to the scientific community as a whole. Explain why this study was important. Then discuss new questions that have emerged from your study. Remember, the more you know, the more you know you don't know! By addressing your original question, you probably uncovered more questions that could be turned into future studies.

Last Paragraph

The last paragraph in the Analysis and Conclusions section should summarize your analysis. The topic sentence should declare the degree to which the results show a relationship between the independent and dependent variable or a difference between the groups. The sentence might begin, "Based on this study, [independent variable] does influence [dependent variable]." (If the research results were unclear, or inconsistent, then the topic sentence might be more like, "Because of the limitations of this study, a connection or lack of connection between [independent variable] and [dependent variable] cannot be made.")

Don't use the word *proves* when talking about this relationship. One experiment does not *prove* anything. Instead, use the word *supports*—for example, "This experiment supports the hypothesis that pressure…." The rest of the paragraph can explain how that final conclusion was made. Another sentence for this final paragraph might be, "Before any strong statement of correlation can be made, additional studies that address the limitations previously mentioned must be conducted." After reading this last paragraph, your reader should know the connections between the variables and the reasons for that correlation or lack of correlation.

Works Cited

See Chapter 10 for a complete discussion of preparing the Works Cited list.

Personal Reflections

In a separate piece of writing (or as a final part of your paper), your teacher may ask you to reflect on the experience of doing your own research and writing the STEM research paper. This is your chance to share how you felt about your experience of being a real scientist. You might use some of the following questions for your personal reflections. The reflections should be real, specific, and honest.

- Why did you do this experiment? What made you curious about this topic?

- Did you find out what you thought you would, or something different?

- How has the question you explored helped you see scientific studies in a new light?

- What about your topic would you have liked to learn more about? Explain.

- Did you have frustrations with the experiment or paper? Explain. How did you overcome them?

- What would you do differently if you were to redo the experiment?

- What advice do you have for students who might be asked to complete their own research studies?

- What experimental topic would you explore if you were to do another research study of this magnitude?

- Why do you think you are required to do this huge research project? What skills did you learn doing this project that will apply to your future?

Abstract (for Oral and Poster Presentations)

An *abstract* is one paragraph that summarizes your entire research project. It is used to help people decide what papers or posters they want to spend time reading. Abstracts include what was studied, how it was studied, the results, and a brief analysis of those results. You will most likely be asked to submit an abstract if you are presenting your results at a research symposium. Your teacher may ask that you submit it with your paper for evaluation. Here are some guidelines for writing an abstract.

Step 1 (one sentence): Describe the project's purpose, using the hypothesis.

Example: The purpose of this research was to determine if increasing the number of hours of light a plant receives also increases leaf width and stem height.

Step 2 (three to five sentences): Describe the methods used.

Example: Three experimental groups and one control group were set up. Each experimental group was exposed to different levels of light—4 hours, 10 hours, and 24 hours. The control group had 12 hours of light. The quantitative measurements—leaf width and stem height—were collected every other day. Data were collected for three weeks.

Step 3 (three to five sentences): Describe the results; be sure to mention the best-performing and least-performing groups and the results of each.

Example: The plants that received 24 hours of light had the most change, with an average of 7 mm of leaf growth and 10 mm of stem growth. The plants that received 4 hours of light had the poorest average growth, with leaf size decreasing by 2 mm and stem growth of 1 mm.

Step 4 (three to five sentences): Conclusion; explain whether or not the hypothesis was supported, and give brief possible explanations for the results.

Example: The hypothesis was not supported because the plants with the least light exposure had the most growth. There may be several explanations for this...

If it makes more sense in terms of your research design, you may want to describe your results and conclusion together in the abstract. For each experimental group, explain what the results were, followed by your explanation of these results. Make a real effort to write very clear explanations.

Preparing the Paper for Submission

If your teacher gave you a grade sheet or rubric for your rough draft, take careful note of the points made there as you write and revise your paper. You'll find examples of both in the appendix section of this book: "Research Paper Grade Sheet" (Appendix C, pp. 197–200) which is a shortened version of the "Research Paper Grading Rubric" (Appendix D, pp. 201–210). Either of these will help focus you as you begin writing a rough draft.

It is always a good idea to schedule a few days between when you finish writing the rough draft and when the rough draft is actually due. Walking away from your writing and then going back with a fresh perspective will increase your ability to make important edits. It is also advisable to have someone else read your paper, especially if this person has written a STEM research paper like the one you have. Your teacher may have you exchange papers with your classmates and use Student Handout #6 (pp. 171—175).

After reading the feedback from your peer editor, you can decide whether to take or reject the advice; it is up to you. If you don't understand some of the editor's comments, talk to him or her directly or seek a second opinion.

Materials to Accompany the Paper

When it is time to submit your final paper, your teacher may ask you to turn in other items together with the paper. These may include the following:

- Approved proposal
- Laboratory notebook
- Access to social bookmarking group
- Note cards, note packet, or access to your online note organization
- Printed internet sources cited in your paper
- Peer-edited and/or rough draft version of the paper

Chapter Question

List and describe the five components of a STEM research paper.

Chapter Applications

Your STEM research paper will have five sections: Introduction, Materials and Methods, Results, Analysis and Conclusions, Works Cited. A poster and/or oral presentation will also require an Abstract. Use this chapter in conjunction with the proposal chapter (Chapter 5) and the MLA documentation chapter (Chapter 10) to keep the writing process moving along. Consider developing a writing schedule for each section of the paper so that you are not cramming too much writing into a single weekend or evening (see Appendix A, pp. 191–192). Also, add some extra time to your schedule so that a classmate can do a peer edit of your paper and you can make adjustments based on that edit before turning in a final copy to your teacher.

If working with others on your STEM research project, agree on tasks for each group member, as you have done at previous stages of the project. Document the discussion in a contract that each member signs, then turn the contract in to your teacher.

The next chapter will help those of you who will be presenting your research to your classmates or to attendees at a symposium or fair.

Reference

Day, R. A., and B. Gastel. 2006. *How to write and publish a scientific paper.* Westport, CT: Greenwood Press.

Recommended Resources

George, M. W. 2009. *The elements of library research: What every student needs to know.* Princeton, NJ: Princeton University Press.

International rules for precollege science research: guidelines for science and engineering fairs. 2010. Retrieved March 16, 2011, from Society for Science and the Public, Intel ISEF document library website: *http://apps.societyforscience. org/isef/rules/rules11.pdf.*

MLA handbook for writers of research papers. 7th ed. 2009. New York: Modern Language Association of America.

Form for Peer Editor of STEM Research Paper

Writer's Name _____

Editor's Name _____

Directions to the Editor:

- You will need a pencil and a highlighter for your editing.

- Before beginning to fill in this form, read the research paper all the way through, without making any marks on it.

- Follow the directions for each question below. Sometimes you will be asked to write something on the research paper; other times you will need to answer on this form.

- If you think that the writer needs to fix, modify, or complete something, write the word FIX next to the number of the question in the left margin on this form.

Paper Setup Questions

1. Circle yes or no for each of the paper setup requirements.

   yes   no   Title is on the first page and paper begins on first page after the title

   yes   no   1 in. margins

   yes   no   Header with name and page number in upper right-hand corner

   yes   no   Figures and tables all fit within the 1 in. margin

   yes   no   Works Cited page is numbered consecutively with the paper

   yes   no   Entire paper is double-spaced.

   yes   no   No double double-spaces (DDs)

Introduction

2. How many paragraphs are in this section? _____

   How many pages are in this section? _____

   Does the section seem to be complete?   yes   no

3. According to the title of the research paper, what are the two elements that are going to be researched? Are these two elements named in the introduction? (They should be.) yes no

Is there at least a paragraph on each of these two elements? yes no

4. Are there any first-person pronouns (*I, me, my,* or *mine*) or second-person pronouns (*you*) used in this section? (There shouldn't be any.) yes no

If there are, circle them.

5. Documentation:

- How many citations (in parentheses) are in the Introduction?

- Do you feel that the writer has provided documentation (a citation) every time it was needed? yes no

  If not, give examples here of a sentences or phrases that seem to need citations:

- Each citation should use one of two MLA citation styles (see Chapter 10 in this book for those two styles: EITHER use author's last name and page number in parenthesis with punctuation after the second parenthesis OR use author's last name in the sentence with the page numbers in parentheses at end of sentence. Underline citations in the Introduction that are not correct.

6. Highlight all sentences that you think are the writer's ideas that help connect the research ideas. (These are good! They make the paper flow.)

Materials and Methods

7. Does the methods section include a photo showing how data were collected? yes no

8. After reading this section, do you think you could you repeat the experiment exactly? yes no

If yes, write an explanation (in praise of the researcher!) regarding the most important parts of the paper that would allow you to repeat the experiment. If no, discuss those specific areas in the paper that you believe need more detail if the study were to be repeated.

170

9. Cross-check the tools and materials in the materials list and in the methods section. Each item should be mentioned in both places. If you find items mentioned in one place but not the other, list them here and where you found them (in the materials list or the methods section):

10. Does the writer explain how extraneous variables were kept constant?   yes   no

11. There should be at least one figure or table in this section. Is there? yes   no

    If yes, write down the sentence from the text that makes reference to the figure or table.

12. Is the hypothesis in future tense?   yes   no

Results

13. How many paragraphs are in this section?_____

    How many pages are in this section?_____

    Does the section seem to be complete?   yes   no

14. Double-underline sentences that highlight trends or patterns in the results. (Highlighting trends or patterns is a good thing! The researcher should not only provide raw data, but also organize data so that trends are clear.

15. Put a single line through any sentence you feel makes a *judgment* or gives an *opinion*. Data should be given in this section without making any inferences. (Inferences about the meaning of the results are not appropriate in this section!)

16. Underline any sentence or paragraph that gives too much day-to-day experimental data.

17. Put a wavy line under sentences that describe irregularities of the results. (Every paper should have wavy lines!)

18. Put a smiley face next to the sentences that describe the group that had the most change. Put another smiley face next to the sentences that describe the group that had the least change.

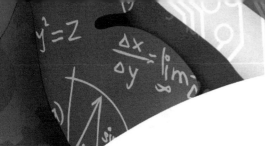

19. Highlight sentences that refer to figures and tables.

Analysis and Conclusions

20. How many paragraphs are in this section? _____

    How many pages are in this section? _____

    Does the section seem to be complete?   yes   no

21. Highlight the sentences that state whether or not the hypothesis was supported.

22. In the first paragraph, write a number in front of each reason that states why the hypothesis was or was not supported.

23. In the body paragraphs of this section, use the numbers you wrote in the first paragraph and put these corresponding numbers in the margin of the paragraph that addresses each reason (*Note:* These reasons should be in the same order they were listed in the first paragraph. Is this true for this paper?   yes   no)

24. Circle the sentences or paragraphs that (a) discuss the control and (b) compare it to the experimental groups.

25. Documentation

    • How many resource citations are in this section? _____

    • Do you feel that the writer provided documentation (citation) every time it was needed?
      yes   no

    If no, give an example of a phrase or sentence that you think needed to be documented as coming from an outside source.

    • Fix any incorrect use of MLA documentation. You might need to refer to Chapter 10 in this book to refresh your memory about the rules for in-text documentation.

26. Put a wavy line under sentences that explain *possible reasons for trends or patterns in the results.*

27. Put a star next to the sentences that explain *how methods or missing constants may have influenced the results.*

Graphical Data Display

28. Go back through the paper and make sure that every figure and table (including any in the appendixes) that appears in the paper is mentioned in sentences within the paper. Put a BIG frowny face next to figures or tables that are not mentioned in the text.

29. Note in the margin if any figures are mentioned out of order from the order that they appear in the paper (e.g., if Figure 3 is mentioned on one page, but Figure 4 and 5 appear in the text before Figure 3 does, that would be an error. Figure 3 has to appear before Figures 4 and 5 are mentioned and appear in the text).

30. Check each graph for proper labeling. Make sure x- and y-axes are labeled correctly, titles are specific, and series (legends) are clear. Make any suggestions for improvement of the graphs right on the paper.

31. List the number and types of figures below. Do they seem appropriate?   yes   no

    Are there too many or not enough?   too many   not enough

Works Cited

32. The entire page is double-spaced. ☺   yes   no

    There are NO double double-spaces. ☺   yes   no

    The entries are in alphabetical order. ☺   yes   no

    If an entry is longer than one line, it uses a hanging indent. ☺   yes   no

    There are at least FIVE entries ☺   yes   no

33. You now have to do some flipping back and forth between the text of the paper and the Works Cited list. THIS MAY TAKE SOME TIME, BUT REMEMBER THAT SOMEONE ELSE IN YOUR CLASS IS DOING THE SAME THING FOR YOUR RESEARCH PAPER. Go through the Introduction and Analysis and Conclusions sections. Every time you find a source in parentheses, flip to the Works Cited list and put a check mark next to that source in the list. If the citation in parentheses is **not** listed in the Works Cited list, circle it. (*Remember:* If there is not an actual author, whatever appears in parentheses should match the first words used in the Works Cited list.)

    Now, collect the following information (this is what a professional editor does):

    • How many sources in parentheses (i.e., the sources you circled above) did not have a corresponding entry in the Works Cited list?_____ Tell the writer that he or she needs to provide the complete reference for the following sources in the Works Cited list:

    • How many entries in the Works Cited list were not mentioned in the paper (i.e., you didn't put a check next to them)? _____ For those entries, the writer has two choices (a) to find legitimate places to insert them in the text or (b) to delete them from the Works Cited list and perhaps put them in a separate list at the end of the paper, called Other Works Consulted. This separate list should follow the same MLA style as the Words Cited list.

    Again, this cross-checking, though tedious, is a tremendous help to the writer. Draw the writer's attention to the fact that his or her paper has inconsistencies between the text and the Words Cited list by writing FIX in the margin to the left.

Spelling and Grammar

34. Mark spelling and grammar errors on the draft.

35. If the word *proves* is found anywhere in the paper, circle it and suggest ways to write the sentence without it.

174

36. Circle any first- or second-person pronouns in the paper as errors and write FIX in the margin of the paper. However, be sure to find out first if the teacher allowed students to use first-person pronouns (*I, me*). In that case, circle only the second-person pronouns (*you*), which are never allowed in a scientific paper. Also, use of *I* and *me* is allowed in the Personal Reflection section of the paper.

37. In your opinion: (1) What is the BEST aspect of this paper? (2) What makes this aspect better than the rest of the paper?

38. In your opinion: (1) What is the part of the paper that needs the most improvement? (2) How could this part be fixed?

Other Comments:

-12-

Presenting the STEM Research Project

Introduction

An important part of conducting research for all STEM professionals is to communicate their study results. This is generally accomplished in two different ways. The first is to write a scientific paper, as outlined in Chapter 11, in order to have the paper published in a STEM or other scientific journal. The other way is to present the research to peers face-to-face at oral presentations and oral poster presentations. This chapter focuses on preparing you for the latter.

Learning Objectives

By the end of the chapter, you will be able to

1. identify the qualities of an oral presentation,

2. explain how visuals can both help and hinder a presentation,

3. describe the components of a poster, and

4. perform an oral presentation of your STEM research project.

Oral Presentations

Even if you are not presenting your research at a research symposium, you may still be expected to present your research to your classmates. Generally, that means an oral presentation—containing visuals—with a question-and-answer session at the end. The presentation may vary in length, depending on

177

your teacher's requirements; student presentations can run from 2 to 15 minutes. See Appendix E, "Oral Presentation Rubric," pages 209–211, for examples of what you will need to consider when preparing your presentation.

If you did your research project with a group, you will need to divide the parts of the presentation among the group members. Consider having each member responsible for his or her own portion, including any corresponding visuals. Then the group can put the parts together before the presentation. Do not have one person do all the preparation and the others do all of the talking on presentation day. Your teacher expects each group member to contribute equally to both the visual and oral parts of the presentation. It is important that each group member verbalize an understanding of the research project. As you have done during previous portions of your research project, you may choose to determine these roles and tasks ahead of time, write them down, and have each member sign the document before turning it in to your teacher.

Parts of the Oral Presentation

Introduction

Begin with one or two sentences to acquaint the audience with your topic. Although you wrote your paper using a formal, impersonal tone, your presentation can include humor, cartoons, current events, a poll of the audience, or before-and-after photos of the entities you studied. *Be creative* and try to *hook* your audience. Your visual (overhead, poster, or electronic presentation) for the introduction should not have a lot of text that your audience has to read. Keep it simple, using graphics with few words.

Hypothesis

Although you do not have to tell the audience your formal hypothesis word for word, you must clearly describe your independent variable and what measurements or correlations you were looking for in your dependent variable. It is fine to use phrases that include the first person, such as, "In my research I was looking to correlate…with…." Again, your visual should be simple, using as few words as possible.

Materials and Methods

Provide just enough detail about your methods so that your audience will understand how you collected data, what extraneous variables you kept constant, and the length of the experiment. In your visual, use photographs you took during your experiment to help tell this story. If your teacher or the symposium or fair officials allow it, bring in parts of your experiment the day of your presentation and integrate them into your explanation of how you collected data. Visuals for the materials and methods portion of the presentation might include photographs of the experimental setup and data collection and a simple table that describes your experimental design, such as the one shown in Table 11.1 (p. 159).

Results

Display and explain your results clearly and succinctly. Carefully choose tables and figures that best highlight the results. When you discuss these graphical data during your presentation, take time to explain *what the graphic shows overall* before launching into the specific results. It takes an audience a moment to orient themselves to what they are viewing, and if you do not take time to help them, they will miss the important points you are making about the table or figure. If you add photographs of the results from the experimental and control groups, you will help your audience even more to connect with your research study.

Analysis and Conclusions

Just as you did in the last paragraph of the analysis and conclusions section in your paper, you need to summarize your conclusions regarding your research. Remind the audience of the connection you were seeking to make, and then describe how confident you are that your research design produced reliable data that either supported, or did not support, the relationship between the independent and dependent variable or a comparison between two groups. Then explain how you reached that conclusion. Be sure to talk about the limitations of the study as well as the factors that may have influenced the results and to speculate on what could have been done to eliminate those factors. Keep in mind that STEM researchers sometimes get results that do not support their original hypothesis. However, they know that they can learn as much from a "failed" experiment as from a "successful" one.

Closure

Close your presentation by making a statement regarding the relationship between the independent variable and the dependent variable. Give a quick summary of why the two are, or are not, linked. Tell your audience how confident you are (you don't have to be 100% confident!) that the results supported this final analysis. Your teacher may also want you to share with members of the audience why they should care about this research study— its importance to the global community and how the knowledge derived from the study can be used.

Asking for Questions

Once you are finished with your presentation, invite questions from your audience.

Preparing for the Oral Presentation

Several Days Ahead

Organize your visuals. Prepare the overheads, poster, or electronic presentation so that they are ready to use for practice. Then, using the visuals, start considering what you will say when each visual is presented. You have spent a lot of time on this project and therefore you should know the content. You can have an outline or note cards or use the note feature in the electronic presentation software when you make your oral presentation, but you should not read from them, only refer to them. The notes should not be written in complete sentences. Write down only the important phrases that will remind you of what you need to say. Finally, think of possible questions that your audience may ask you and figure out how you might answer those questions. Review your background research notes so that you will be prepared!

Practicing the Presentation

Putting your visuals and notes together is the first step of preparation. You should practice your presentation out loud, several times, using your visuals and your notes. Your first or second time, focus on effectively delivering the ideas. Make sure you know what you will say and be sure that the visuals are a crucial part of the presentation. Point to places on the tables and figures as you discuss them. Pick up and use parts of your experiment to make a point or to demonstrate a method you used.

Once you feel confident that you can effectively communicate information about your STEM research project, work on polishing the presentation to make sure the audience hears what you have to say. While practicing out loud, work on clear pronunciation (try not to mumble, even if you are nervous), breathing, voice fluctuation, enthusiasm, and pauses. Pauses are an important part of public speaking. But all pauses are not equally welcome. People new to public speaking often fill quiet moments with *um…, OK…, like…,* or *so….* It is better to let the silence be silence (hard as that may be) then to use verbal fillers.

Most likely you will have to stand while giving your presentation. Therefore, stand while you practice. Get comfortable with this position. Work on posture and eye contact. Practice looking at *all* parts of the audience, sweeping your eyes to both sides of the room. Figure out what to do with your hands during the presentation. Practice how you will conclude your presentation. Find a natural and upbeat ending statement that flows from what you have been talking about—perhaps say how exciting one particular aspect of the learning process has been for you or how you look forward to your research being applied to new scientific questions. Don't end saying, "That's all I have to say" or "I guess I'm done."

The Day of the Presentation

Now that the day is here, there are several things you can do to help it go smoothly. First, be sure to dress appropriately—you are giving a formal presentation of your time-consuming and important work. Don't wear jeans and a T-shirt. Be sure all your materials are ready for the presentation. Have your visuals well protected and packed for easy transport. If you have electronic visuals, save them in several locations to ensure you can gain access to them at presentation time. Consider saving them by e-mailing the file to yourself, posting it to your personal server space at school, and saving it externally so that it can be directly connected to a computer and ready to use. If your teacher doesn't mind, bring a water bottle to have available during your presentation, in case your mouth is dry. (Also, taking a sip of water will help you fill in a pause.) Although you might like to chew gum during the presentation to calm yourself or prevent dry mouth, *don't*. It is distracting to the people in the audience and they are less likely to take you seriously.

Once you are in front of the room and begin to talk, go for it—just as you practiced. If something goes wrong—perhaps your experiment falls on the floor or you have trouble accessing your presentation on the computer—do not overreact. Breathe deeply, fix the problem without comment, and continue. The more calmly you handle the problem, the more confident you will feel. Once you have completed your presentation, take questions from the audience. Answer questions as truthfully as you can. If you know the answer to a question, confidently answer it. However, do tell your audience if your answer is your opinion, something you read in your background research, or a result from your study. If you don't know the answer, admit it. Don't make up answers. However, feel free to give your best guess, as long you say that that is what you are doing. Here are some possible responses to audience questions:

> *I thought of that question as well, but couldn't find any research to answer it. However, with what I learned in my experiment, I think that maybe A does affect B because....*
>
> *Yes, you're right. I didn't consider that as a possible variable that I would need to control. If I did the experiment again, I would probably solve that problem by....*

When you are finished taking questions, smile, take a seat, and bask in a job well done.

Responsibilities as an Audience Member

You have an important responsibility to be an active listener during the presentations by your classmates (even if your presentation is next in line and that's all you can think about!). Below are some suggestions for being a good audience member.

First, listen and take notes. Take notes on the important points of each presenter, such as the variables he or she was testing, results, and conclusions, taking special note of the experimental design. See Appendix B, "Research Presentations Observation Sheet," for help with organizing your notes (pp. 195–196).

Second, be prepared to ask questions. Don't ask superficial questions such as, "Did you like doing this project?" or other questions that could be asked about any project. Instead, really listen to the experimental design of the presenter and be looking for ways to improve on the study he or she conducted. Feel free to ask clarifying questions about something you did not understand. Here are some questions to have in mind when listening to a presentation. (*Hint:* These are questions teachers or science fair judges may ask when assessing papers and presentations.)

- Did the researcher keep experimental groups and control groups the same?

- Were there extraneous variables that the researcher overlooked?

- Were there flaws in how the experiment was set up?

- Did the methods affect the results in ways not addressed by the researcher?

- Were the results explained well?

- Did the researcher explain possible reasons for flukes or inconsistent data?

- Were the researcher's conclusions consistent with the data that were presented?

When asking questions, be respectful. Just because a student doesn't mention a specific aspect of his or her research in the presentation, does not mean he or she has not thought of it. The oral presentation is only a few minutes, and much has to be left out. Therefore ask your questions in a tactful, respectful way: "Is it possible that while the experiment was being set up, A affected B and this influenced the results somehow?"

You can be proud of yourself if you are able to summarize other people's research and ask quality questions. That means that you not only learned good scientific research methods while doing your own project but that you are able to apply those methods to other projects as well. That application skill is a valuable one that is difficult to teach as well as to learn.

Oral Poster Presentations

Poster presentations differ from other presentations in that the audience members move around to the presenters, each of whom is standing by a poster that summarizes his or her research study. This allows many researchers to share their studies at one time and allows audience members to focus on specific topics that interest them. In preparation for the poster presentation, you will need to create a poster and prepare a short, two-minute speech about your research study.

You'll probably be able to use and modify your paper for the poster. However, you'll have to do more than just print out the paper and stick it to a poster board. Carefully read the suggestions below.

Creating the Poster

Follow the specific requirements given to you by your teacher and/or the institution that is hosting the poster presentation or symposium you will be attending. Rubrics, grade sheets, or requirements checklists are often provided ahead of time to help student researchers prepare for a symposium. Be sure that all the elements that will be judged are present in your poster.

Poster Design Tips

Posters will most likely be placed on easels or tables. Foam board, mat board, or strong refrigerator cardboard are common materials for posters. Tri-fold display boards, available at most craft and hobby stores, are commonly requested. Poster sizes are large, ranging from 36 in. × 36 in. to 40 in. × 60 in.; check with the people who are running your poster presentation or symposium before making a purchase. Most do not want the flimsy poster boards readily available at drug stores.

A person viewing the poster should have no trouble reading it from a distance of 5 ft. All words on the poster should be typed; choose font styles and sizes appropriately. Use color, illustrations, photographs, figures, tables, and other visuals along with text to make it more inviting. Consider using arrows to indicate the flow and organization of the poster. Craft and hobby stores have all kinds of poster supplies.

Poster Components

These are the poster components that are most commonly requested for research symposium events and other poster presentations. Check for specific requirements. Also refer to Chapter 11 for greater detail on what information goes in each section.

Descriptive Title

The title should describe the entity and two variables studied. It should be in large print and easy to read.

Hypothesis

State your hypothesis or research question word for word. It should be placed in a prominent place on your poster.

Introduction

Describe your background research. As in your paper, cite the resources you used and provide a Works Cited list.

Materials and Methods

You do not need to list the materials, but make sure they are all mentioned in your description of how you set up the experiment. You have limited space, so briefly describe what you did, how you collected both quantitative and qualitative data, and how often. Add enlarged photographs of your experimental setup to help clarify.

Results

Describe and display your data. Using photographs, figures, and tables as well as written text, show the results of your experiment. Be sure that the figure and table numbers match the numbers you use in the written text.

Conclusions

In the first sentence of this section, restate the hypothesis or research question and say whether or not a clear connection can be made between your two variables. Then use background research (appropriately documented) to explain *why* the results may have occurred. Include procedures that may have influenced the results.

Works Cited

Include an MLA-style Works Cited list (unless the rules for the symposium require that you use the reference style of the American Psychological Association [APA] or another style).

Other Works Consulted (optional)

Here is the place to list those resources that you used in your background research but didn't cite in your paper. Whether or not to include this list is up to you.

Abstract

The abstract is a paragraph that briefly describes the experiment. The abstract is what people can read if they don't have time to read all parts of your poster. It should be short, clear, and to the point. It describes your hypothesis, what you researched, the results, and an explanation of the results. (*Note:* Your abstract may or may not actually be on the poster. It is often submitted at the time you register your poster for a symposium. The symposium officials might put the abstracts from all the presenters into a booklet or put them online before the symposium so that people can browse the topics in advance.)

Preparing for the Poster Symposium

Many of the same suggestions that were described earlier for oral presentations apply to poster presentations as well. You should dress professionally, have water, and make sure that your poster is complete. You might consider bringing emergency supplies, like a glue stick, in case you need to do repairs. Most likely, you will be assigned a place in the room to put your poster as well as a specific presentation time. You are required to stand by your poster during that entire time (anywhere between one and three hours) to answer the questions of viewers, including judges, who are circulating around the room.

Stand by your poster, and as people linger, start a conversation with them. Ask if they would like to hear about your research project. Be prepared to give a short (two-minute) overview; don't be surprised, however, if people interrupt you to ask a question.

Audience members at poster symposiums and presentations include professors, graduate students, undergraduate students, teachers, and other students your age who have done research. It may be that you will not know which person is a judge; therefore, you must put on your best show for every person you talk to. If you don't know the answer to a question, don't pretend you do. Say, "I'm not sure. I didn't come across anything to answer that in the research I did." Keep eye contact as you say this (don't look down; you have nothing to be embarrassed about!). If you're answering a question but are not sure if your answer is accurate, preface it with a statement such as, "I'm not sure, but to the best of my knowledge, this…is a possibility."

Being a Poster Symposium Audience Member

During the times in the day when you are not presenting, you might want to circulate and visit other students' posters. If the other students don't engage you in conversation, take the first step and ask them about their research. Then follow up with quality questions. Mentally, try to find weaknesses in their experimental designs, such as aspects they did not keep constant. When asking about these possible weaknesses, use respectful questions such as, "Did you consider the possibility of…" or "How did you control…." Also, focus questions on whether presenters analyzed their findings. Did they come up with explanations that directly relate to the data they collected? Once you have a good idea of a person's experiment, thank him or her and move to another poster. Remember, these presenters are just as nervous as you were when you were being questioned. It is helpful for everyone if you can put them at ease.

Chapter Questions

1. What are the qualities of a good oral presentation?

2. How can visuals both help and hinder a presentation?

3. Describe the components of a poster.

Chapter Applications

You should prepare a day-by-day schedule of what you need to do for your presentation, working back from the date of the presentation. Give yourself plenty of time to do the tasks listed in this chapter, especially practicing your presentation out loud. When you present your results orally and visually, you are going through the same process as that of professional STEM researchers.

In Chapter 1, I described the scientific research process as "messy." Now that you have been through the process yourself, you may see why the linear analogy of the staircase is not realistic. Instead of being frustrated by the fact that the research process is not always clear-cut and uncomplicated, I hope you can see the power of being flexible. Scientists always have a well-thought-out plan before beginning any research study but must remain open to what happens within their research and make appropriate changes as they go. The more you know, the more you realize you don't know.

You have completed a STEM research experiment by doing extensive background research, designing an experiment, collecting data by keeping it organized in a laboratory notebook, analyzing data, writing a scientific paper, and presenting the results to your peers. It's official! You have experienced STEM research in the truest sense of the term. The next time you hear a news report that begins, "The latest research shows…," you will have a firsthand understanding of all the time and effort that went behind the words.

Appendixes

Appendix A

Research Project Due Dates Checklist

Name(s):_____

Research Topic: _____

Research Project Title: _____

Directions: This checklist will help you keep track of upcoming due dates and what you have turned in. For some pieces of work, your teacher might ask you to do several drafts before giving his or her approval for you to go forward. I call this part of the process a **DUA**—meaning a piece of work that you need to **Do Until Accepted.** Try to keep up with the ORIGINAL due dates, but remember—it often takes multiple drafts to get an assignment approved. (You will note that there are four **DUAs** in the stages below.)

Stages to Be Completed	Due Date	Deadline for Final Approval by Teacher (for DUAs only)	Completed
Complete "Student Handout #1: Focusing Preliminary Research Ideas"			
Complete "Student Handout #2: Research Design Table" **(DUA)**			
Complete "Student Handout #3: Background Research Questions"			
Turn in Final Draft of Background Research Questions **(DUA)**			
Show Notes You Have Taken to Your Teacher			
Complete "Student Handout #4: Practicing Writing Hypotheses"			
Turn in Research Proposal **(DUA)**			
Organize Laboratory Notebook (including empty tables)			
Start Your Experiment			

Stages to Be Completed	Due Date	Deadline for Final Approval by Teacher (for DUAs only)	Completed
Provide Teacher With Evidence of Your Data Keeping • Laboratory Notebook • Photographs of Experimental Setup			
End Your Experiment			
Organize Data Into Tables and Graphs **(DUA)**			
Complete "Student Handout #5: Interpreting Graphical and Statistical Data: A Peer Editing Exercise"			
Turn in Rough Draft of Research Paper			
Complete "Student Handout #6: "Form for Peer Editor of STEM Research Paper"			
Submit Final Research Paper			
Deliver Classroom Presentation			
Participate as Audience Member in Classroom Presentations by Other Students			
Complete Poster (if you are creating a poster in addition to or in place of the research paper)			
Participate in Pseudo-Symposium Presentation			
Participate in Real Symposium Presentation			
Give Your Teacher Evidence (Notes) of Symposium Participation			

Appendix B

Research Presentations Observation Sheet

(For students in the audience and/or the teacher)

Name: _____

Date: _____

Name(s) of Student Presenter(s)	Observation Notes: Include the topic of the research project, the main ideas of the presentation, and any questions you have.

Appendix C

Research Paper Grade Sheet

(For the teacher)

Student Name(s): _____

Research Project Title: _____

Date: _____

Research Paper Sections	Points
Introduction	(Possible # of Points: 12)
❏ Description of the problem provides critical background on the need for and/or importance of the current research project.	
❏ Background on entity's qualities (and care and safety, if applicable) is sufficient and well described.	
❏ Discussion of previous scientific research on the Independent Variable (IV) supports the manipulation of this variable and indicates that manipulation will help address the hypothesis or research question.	
❏ Discussion of previous scientific research on the Dependent Variable (DV) indicates that it is a good variable to measure or observe in response to the IV. Known associations with the IV are described.	
❏ Options for quantitative data collection and measurement are well described. Best methods are justified.	
❏ Options for qualitative data collection and measurement are well described. Best methods are justified.	# of Points Assigned:

Research Paper Section (continued)

Points

Materials and Methods

(Possible # of Points: 14)

❏ Description of how the experiment was prepared for data collection is detailed enough that another individual could replicate the experiment. Description of how quantitative data were collected is detailed enough for someone to replicate the research.

❏ Data collected are appropriate for answering the research question.

❏ Description of how qualitative data were collected is detailed enough for someone to replicate the research.

❏ Data collected are appropriate for answering the research question.

❏ Photographs of experimental setup and data collection are included, referred to within the text, and increase the reader's understanding of how the research was conducted.

❏ Regular tasks that occurred throughout the experiment are thoroughly described, including the tools used to complete them.

❏ Surveys, large assessments, or ISEF paperwork are included, labeled, and referenced correctly in the paper.

❏ Research design is well described and makes clear the difference between control and experimental groups.

❏ Extraneous variables are described. Explanations of how they were kept constant or monitored are given.

# of Points Assigned:

Hypothesis

(Possible # of Points: 4)

❏ Hypothesis is written as a testable statement or question; it includes IV, DV, and predictions that can be supported or rejected.

# of Points Assigned:

Results

(Possible # of Points: 16)

❏ This section clearly describes how raw data were calculated and organized for analysis.

❏ Data are logically organized either by groups or by type of data. The organization chosen clearly highlights the groups and/or trials that had the most and least change.

❏ All quantitative data were accurately calculated and are appropriately represented in the text.

❏ All qualitative data are objectively described and are correctly represented in the text.

❏ Direct statements compare how each of the groups compares to one another.

❏ The text specifically refers to aspects of the graphs and tables that help highlight trends or patterns.

# of Points Assigned:

Research Paper Section (continued)

Graphical Representation of Data	(Possible # of Points: 8)
❏ All figures and tables are appropriate for the type of data and best allow for comparison among groups.	
❏ The number of figures and tables is appropriate for proper interpretation of the results.	
❏ All calculations are mathematically correct and are accurately represented in graphical form.	
❏ Titles of figures and tables accurately and completely depict what is contained within the figure or table. Labels on graphs allow readers to interpret the graphs without having to read the text. The labels include units and axes or columns and rows.	# of Points Assigned:
Analysis and Conclusions	(Possible # of Points: 18)
❏ This section of the paper contains a statement that accurately indicates which data support the hypothesis and which data do not.	
❏ Paragraphs within this section are logically organized and connected, explaining data and results in the context of the hypothesis.	
❏ Conclusions regarding quantitative data are logical, are based on study data, and have resources to support the position.	
❏ Conclusions regarding qualitative data are logical, are based on study data, and have resources to support the position.	
❏ Comparative explanations of the experimental and control groups are logical and have resources to support the position.	
❏ Data that do not follow trends (outliers) are thoroughly and specifically explained.	
❏ This section addresses the possibility of data collection errors and/or ways the research design may have introduced limitations.	
❏ Limitations are correctly identified and described. Suggestions are provided for how future studies could improve on the study.	
❏ With appropriate certainty, the student researcher draws conclusions about the relationship between the IV and DV.	
❏ The student researcher connects the research study to possible real-world applications and provides plausible ideas for future studies.	# of Points Assigned:

MLA Documentation

❏ In-text citations for ideas or quotations that came from other sources use either (a) Author Name at End of Sentence style or (b) Author Name Inside of Sentence style (see Chapter 10 in this book).	(Possible # of Points: 12)
❏ All citations within the paper can be found in the student researcher's notes (shows strong evidence that paper was not copied directly from resources).	
❏ All citations within the paper are listed in Works Cited and all references in Works Cited are cited in the paper.	
❏ All entries in Works Cited are correctly formatted and are in alphabetical order and from reliable sources.	# of Points Assigned:

Spelling and Mechanics

❏ Paper contains proper spelling throughout.	(Possible # of Points: 8)
❏ Paper contains proper grammar throughout.	
❏ All paragraphs have well-written topic sentences that accurately describe the content within the paragraph.	# of Points Assigned:
❏ Transitions are used when comparing data and when shifting to new ideas.	

Scientific Writing

❏ Proper tense and voice (as specified by the teacher) are used throughout.	(Possible # of Points: 6)
❏ Entire paper is written using formal grammar, in clear, focused language.	
	# of Points Assigned:

Page Setup

❏ Margins are 1 in. around entire paper; opening page and "headers" are set up correctly; paper is double-spaced throughout.	(Possible # of Points: 2)
	# of Points Assigned:

(Possible Number of Points: 100) **Points Earned:**	

Appendix D

Research Paper Grading Rubric

(For the teacher)

Student Name(s): _____

Research Project Title: _____

INTRODUCTION _____ / 12 points possible

Problem Described

2	1	0
Description of the problem provides critical background on the need for and importance of the current research study.	Description of problem is weak and/or poorly documented.	Problem is not described.

Research on Entity Studied

2	1	0
Sufficient background on entity's qualities (and care and safety, if applicable) are well described.	Background research on entity is weak and/or incomplete.	Background research on entity is absent.

Research on Independent Variable (IV)

2	1	0
Background research on IV meets both of the criteria listed: Research (1) supports the manipulation of this variable and (2) indicates that manipulation will address the hypothesis or research question).	Background research on IV meets only one of the two criteria.	Meets neither criteria.

Research on Dependent Variable(s)

2	1	0
Background research on DV meets both of the criteria listed: (1) DV is a good variable to measure or observe in response to IV; and (2) known associations of DV with IV are described.	Background research on DV meets only one of the two criteria.	Meets neither criteria.

Research on Quantitative Methods

2	1	0
Quantitative research meets both of the criteria listed: (1) Options for quantitative data collection and measure are well described; and (2) best methods are justified.	Quantitative research meets only one of the two criteria.	Meets neither criteria.

Research on Qualitative Methods

2	1	0
Qualitative research meets both of the criteria listed: (1) Options for qualitative data collection and measurement are well described; and (2) best methods are justified.	Qualitative research meets only one of the criteria.	Meets neither criteria.

MATERIALS AND METHODS _____ / 14 points possible

Experimental Setup

2	1	0
Description of how the experiment was prepared for data collection is detailed enough that another individual could replicate it.	Description missing small details that would make it easier for someone to replicate the research.	Description is missing critical details necessary for someone to be able to replicate the research.

Quantitative Data Collection

3	2	1
Description of how quantitative data were collected is detailed enough for someone to replicate the research, and data collected is appropriate for answering the research question.	Description missing small details that would make it easier for someone to replicate the research.	Description is missing critical details necessary for someone to be able to replicate the research, or data collected does not help answer the research question.

Qualitative Data Collection

3	2	1
Description of how qualitative data were collected is detailed enough for someone to replicate the research, and data collected is appropriate for answering the research question.	Description missing small details that would make it easier for someone to replicate the research.	Description is missing critical details necessary for someone to be able to replicate the research, or data collected does not help answer the research question.

Photographs (if applicable)

1	1 (Not Applicable)	0
Photographs of experimental setup and data collection (1) are included, (2) are referred to within the text, and (3) increase the reader's understanding of how the research was conducted.	Not applicable, no reduction of points.	Missing one or more of the criteria (see column at right).

Experimental Tasks (hourly, daily, weekly)

1	0
Regular tasks that occurred throughout the experiment are thoroughly described. Tools used to complete the task are also thoroughly described.	Tasks and/or tools only briefly described or missing completely.

Additional Materials Provided in Appendix (if applicable)

1	1 (Not Applicable)	0
Surveys, large assessments, or ISEF paperwork are (1) included, (2) labeled, and (3) referenced correctly in the paper.	Not applicable, no reduction of points.	Missing one or more of the criteria (see column at right).

Research Design

1	0
Appropriately describes research design, making clear the difference between control and experimental groups.	Research design not clearly explained.

Extraneous Variables

2	1	0
Properly addresses extraneous variables and explains how they would be kept constant or monitored.	Briefly addresses extraneous variables but does not sufficiently explain how they would be kept constant or monitored.	Does not address extraneous variables.

HYPOTHESIS _____ / 2 points possible

2	1	0
Meets two or more of the criteria: Hypothesis is (1) written as a testable statement or question; (2) includes IV and DV; and (3) includes predictions that can be supported or rejected.	Missing one of the criteria.	Missing all three criteria.

DATA AND RESULTS _____ / 16 points possible

Preparation of Data

2	1	0
Clearly describes how raw data were calculated and organized for analysis.	Briefly explains how raw data were organized for analysis.	Does not explain how raw data were organized for analysis.

Section Organization

2	1	0
Data logically organized either by groups or by type of data. Organization chosen clearly highlights best and worst performing groups/trials.	Section inconsistently organized.	Section lacks paragraph organization and makes reading the paper difficult.

Quantitative Data

4	2–3	1
All quantitative data were accurately calculated and appropriately represented in the text.	Most quantitative data were accurately calculated and appropriately represented in the text.	Major errors in the calculation of quantitative data or in the representation of the data.

Qualitative Data

4	2–3	1
All qualitative data were objectively described and correctly represented in the text.	Most qualitative data were objectively described and correctly represented in the text.	Major errors in the description of qualitative data.

Experimental and Control Group Comparisons

2	1	0
Direct statements compare how each of the groups compares to one another.	General references compare each of the groups to one another.	No statements compare the groups to one another.

Connection Between Narrative and Graphical Representation of Data

2	1	0
Narrative includes essential references to specific aspects of figures and tables that help highlight trends or patterns.	References to tables and graphs are only general and do not refer to specific trends or patterns shown in the tables and graphs.	References to tables and figures are missing.

GRAPHICAL REPRESENTATION OF DATA _____ / 9 points possible

Choice of Graphical Data

3	2	1
All figures and tables used are appropriate for the type of data and best allow for comparison among groups.	One figure or table is inappropriate for the type of data.	Two or more figures or tables used are inappropriate for the type of data.

Number of Graphical Data

2	1	0
Number of figures and tables is appropriate for proper interpretation of the results.	Interpretation of results would be improved with the addition/ subtraction of tables/graphs.	Number of figures and tables is inappropriate (either too many or not enough).

Calculations Within Graphical Data

2	1	0
All figures and tables meet the criteria: (1) All calculations are mathematically correct; and (2) all calculations are accurately represented in graphical form.	One figure or table does not meet the criteria.	Two or more figures or tables do not meet both criteria.

Graphical Data Titles and Labels

2	1	0
Meets two or more of the criteria: (1) Titles accurately and completely convey what is contained within the table or figure; (2) labels on graphs allow readers to interpret the graphs without having to read the text; and (3) labels include units and axes or columns and rows.	One figure or table is missing criteria.	Missing all three criteria.

ANALYSIS AND CONCLUSIONS _____ / 19 points possible

Do Data Support Hypothesis/Research Question?

1	0
Statement accurately indicates which data support the hypothesis and which data do not.	Incorrect, incomplete, or weak statement connecting data to hypothesis.

Section Organization

2	1	0
Paragraphs in this section are logically organized and connected. They explain data and results in the context of the hypothesis.	This section is inconsistently organized.	This section lacks paragraph organization and makes reading the paper difficult.

Quantitative Data

4	2–3	1
Conclusions regarding quantitative data meet two or more of the criteria: Conclusions (1) are logical, (2) are based on study data, and (3) have resources to support the position.	Conclusions regarding quantitative data are missing one of the criteria.	Missing all three criteria.

Qualitative Data

4	2–3	1
Conclusions regarding qualitative data meet two or more of the criteria: Conclusions (1) are logical, (2) are based on study data, and (3) have resources to support the position.	Conclusions regarding qualitative data are missing one of the criteria.	Missing all three criteria.

Experimental and Control Group Comparisons

1	0
Comparative explanations of the experimental and control groups are logical and have resources to support the explanations.	Weak or no comparative explanations are included.

Outlier Data (if applicable)

1	1 (Not Applicable)	0
Data that do not follow trends are thoroughly and specifically explained.	Not applicable, no reduction of points.	Data that do not follow trends are briefly and generally explained or are not addressed.

Researcher and Research Design Error

2	1	0
Appropriately addresses the possibility of data collection errors and/or ways the research design may have introduced limitations.	Briefly admits the possibility of data collection and research design errors but does not explain why those errors put limitations on the study results.	Researcher and/or research design error is not addressed.

Study Limitations and Suggested Improvements

2	1	0
Correctly identifies and describes limitations, and provides suggestions for how future studies could improve on the study.	Either identifies and describes limitations, or provides suggestions for how future studies could improve on the study.	Neither identifies limitations, nor provides suggestions for how future studies could improve on the study.

Statement of Conclusion: Relationship of IV and DV

1	0
With appropriate certainty, draws conclusions about the relationship between the IV and DV.	Makes incorrect conclusions about the relationship between the IV and the DV or lacks any conclusion statements at all.

Application and Future Studies

1	0
Correctly connects the research study to possible real-world applications and provides plausible ideas for future studies.	Weak or absent connection of research to real-world applications and future studies.

PERSONAL REFLECTION _____ / 4 points possible

Your View of Science

2	1	0
In detail, describes how your view of science has or has not been altered by completing this project.	Briefly describes how your view of science has or has not been altered by completing this project.	Does not address your view of science.

With What You Now Know

2	1	0
In detail, describes what you would do differently if you were to do the project again.	Briefly describes what you would do differently if you were to do the project again.	Does not address changes to implement next time.

DOCUMENTATION _____ / 8 points possible

Parenthetical Documentation (That is, for in-text citations for ideas or quotations that came from other sources, the writer has used either (a) "Author Name at End of Sentence" style or (b) "Author Name Inside of Sentence" style [see Chapter 10 in this book for explanation].)

2	1	0
Correctly formatted parentheses follow all statements that came from another source.	Two to three errors in parenthetical documentation.	More than three errors in parenthetical documentation.

Correlation of Notes and Paper

2	1	0
All citations in paper can be found in notes. (Shows strong evidence that paper was not written copying directly from resources.)	Two to three errors.	More than three errors.

Correlation of Parenthetical Citations With Works Cited

2	1	0
All citations in the paper are listed in Works Cited and vice versa.	Two to three errors.	More than three errors.

Works Cited

2	1	0
All entries are correctly formatted, in alphabetical order, and from reliable sources.	Two to three errors in Works Cited.	More than three errors in Works Cited.

SPELLING AND MECHANICS _____ / 8 points possible

Spelling

2	1	0
Paper contains proper spelling throughout.	One to two errors in spelling	More than two errors in spelling.

Grammar

2	1	0
Paper contains proper grammar throughout.	One to two errors in grammar.	Three or more errors in grammar.

Topic Sentences

2	1	0
All paragraphs meet the two criteria: Topic sentences (1) are well-written and (2) describe the content within the upcoming paragraph.	One to two paragraphs are missing criteria.	Three or more paragraphs are missing both criteria.

Transitions

2	1	0
Transitions used appropriately when comparing data and when shifting to new ideas.	Inconsistent use of transitions.	Poor use of transitions makes reading the paper difficult.

SCIENTIFIC WRITING _____ / 6 points possible

Tense and Voice
(Grading will depend on teacher's directions regarding use of past or present tense and active or passive voice.)

2	1	0
Entire paper is in past tense and passive voice.	One to two errors in tense and/or voice.	Three or more errors in tense and/or voice.

Person

(Grading depends on whether teacher has permitted use of first person ("I"). There should be no use of second person ("you").

2	1	0
Entire paper is written without the use of first or second person.	One to two errors in person.	Three or more errors in person.

Scientific Language

2	1	0
Entire paper written using formal grammar, with clear, focused language.	Paper lacks either formal grammar or focused language.	Paper lacks both formal grammar and focused language.

PAPER SETUP _____ / 2 points possible

2	1	0
Meets the three criteria: (1) Margins are 1 in. around entire paper; (2) opening page and "headers" are set up correctly; and (3) paper is double-spaced throughout.	Missing one of the three criteria.	Missing all three criteria.

Points Earned _____

Points Possible _____ 100 _____

Instructor's Comments:

_____

_____

_____

_____

_____

_____

_____

_____

_____

_____

Appendix E

Oral Presentation Rubric

(For use by teacher—or by students if they are evaluating the presentations of their classmates)

Name(s) _____

Research Topic _____

Research Project Title _____

DELIVERY _____ / 15 points possible

Body Posture, Gestures, and Movement

3	2	1
Posture, gestures, and/or movement are helpful and not distracting. Stands straight and uses purposeful gestures and movements.	Unnatural posture, gestures, and/or movement are noticeable but do not detract from the presentation.	Posture, gestures, and/or movement interfere with or detract from the presentation.

Eye Contact

3	2	1
Maintains consistent eye contact with entire audience.	Maintains eye contact with most of audience most of the time.	Has very little or no eye contact with audience.

Volume and Voice Projection

3	2	1
Speaks loudly and comfortably to be heard by the entire audience.	Speaks loudly enough to be heard by most audience members.	Not heard by everyone; speaks softly causing some audience discomfort.

Rate and Intonation

3	2	1
Varies rate of speech and changes voice tone for natural effect throughout the presentation.	Varies either rate OR tone.	Varies neither rate nor tone.

Pauses

3	2	1
Uses pauses appropriately, not interrupting quiet moments.	Occasional fillers are used but do not detract from the presentation.	Verbal fillers like *um, ah,* or *OK* interfere with and detract from the presentation.

CONTENT _____ / 35 points possible

Introduction

3	2	1
Meets all of the criteria: Introduction provides (1) a brief but clear description of the entity studied, (2) brief but clear descriptions of the IV and DV, and (3) a brief but clear statement of hypothesis or research question.	Missing one criterion.	Missing all three criteria.

Body and Organizational Pattern

3	2	1
The speaker follows a clear and logical organizational pattern.	The speaker attempts to use a pattern of organization.	The speaker is unorganized.

Research Design

3	2	1
Meets both criteria: Speaker offers (1) a clear description of the difference between the control and experimental groups and (2) clear description of how extraneous variables were kept constant or monitored.	Missing one criterion.	Missing the two criteria.

Methods

4	2–3	1
Thoroughly explains how the research was completed and how quantitative and qualitative data were collected.	Briefly explains how the research was completed and how data were collected.	Does not provide clear details about methods, causing confusion in audience once discussion of results begins.

Data and Results: Quantitative and Qualitative Data

5–6	3–4	1–2
Both quantitative and qualitative data are explained sufficiently and logically in context with the hypothesis or research question.	Fair explanation of data and/or attempts to organize results in context with the research question.	Poor explanation of data and/or do not help audience connect results to research question.

Visual Aids: Graphical Representation of Data

3	2	1
Visuals were vital to the presentation. Choice of visuals was appropriate to show data and results as well as the analysis and conclusion.	Visuals poorly used. Choice of figures was not appropriately used to drive discussion of data and results or the analysis and conclusion.	Visuals either detracted from the presentation or were not strong enough to contribute to discussion of data and results or the analysis and conclusion.

210

Analysis and Conclusion

5–6	3–4	1–2
Meets two or more of the criteria: Presenter offers (1) clear conclusions based on data, (2) explanations of why the results may have occurred, and (3) explanations of how the experiment could be improved.	Missing two criteria.	Missing all three criteria.

Closure: Statement of Relationship of IV and DV

3	2	1
Meets both criteria: Presenter provides (1) a clear statement of whether or not the research questions were supported by the data and (2) a statement of conclusions that can be drawn about the relationship between the IV and DV.	Missing one criterion.	Missing both criteria.

Questions

4	2–3	1
Speaker confidently answers questions from the audience.	Speaker is able to answer most questions asked by the audience.	Speaker is unable to answer questions from the audience.

Points Earned _____

Points Possible _____50_____

Teacher's Comments:

_____

_____

_____

_____

_____

_____

_____

_____

_____

Appendix F

Judge's Score Sheet for STEM Research Projects

Session _____ Poster # and Title _____

School _____ Teacher _____

INTRODUCTION _____/8 points possible

		Disagree			Agree
1.	Introduction includes strong background research on entity studied and on dependent and independent variables.	1	2	3	4
2.	Documentation is prevalent throughout the introduction.	1	2	3	4

METHODS AND HYPOTHESIS _____/12 points possible

3.	Methods chosen (research design) appropriately address the hypothesis or research question.	1	2	3	4
4.	Research design is described and makes clear the difference between control and experimental groups.	1	2	3	4
5.	How extraneous variables were kept constant or monitored is explained.	1	2	3	4

RESULTS _____/12 points possible

6.	Data are calculated and organized appropriately.	1	2	3	4
7.	Data are accurately represented in the text.	1	2	3	4
8.	Graphic representation of the results is appropriate for data type and allows for comparison between the groups.	1	2	3	4

CONCLUSIONS _____/24 points possible

9.	Experiment conclusions are logical and based on study data.	1	2	3	4
10.	Results are explained in the context of the hypothesis.	1	2	3	4
11.	The possibility of data collection errors and/or ways the research design may have introduced limitations is well explained.	1	2	3	4

12. Conclusions about the relationship between the independent and dependent variable are drawn with appropriate certainty.   1   2   3   4

13. Results are correctly connected to possible real-world applications and/or provide plausible ideas for future studies.   1   2   3   4

14. Documentation is prevalent throughout the conclusion section.   1   2   3   4

GRAMMAR, USAGE, AND MECHANICS _____/4 points possible

15. Scientific writing in clear focused language is prevalent over casual informal writing.   1   2   3   4

REFERENCES _____/12 points possible

		Disagree			Agree
16.	References are appropriately cited within the text.	1	2	3	4
17.	Citations within the text can easily be found in the reference list.	1	2	3	4
18.	Various types of references are used and the references are from reliable sources.	1	2	3	4

VISUAL APPEAL _____/8 points possible

19. The visual content is presented in a logical format.   1   2   3   4

20. Poster is professional in the use of colors, fonts, and graphics.   1   2   3   4

ORAL PRESENTATION _____/20 points possible

21. The student speaks clearly (maintains eye contact, with few "ums") and uses appropriate scientific language.   1   2   3   4

22. The student is well prepared, uses complete sentences, and pronounces words correctly.   1   2   3   4

23. The student is knowledgeable about the information he or she is presenting.   1   2   3   4

24. The student summarizes all the key points of his or her topic.   1   2   3   4

25. The student accurately answers all questions.   1   2   3   4

Total _____/100

Judge's comments *(will be used for tiebreaker):*

_____

_____

NOTE: Posters with plagiarized content will be disqualified.

INDEX

216

218